Contents

Learning how to influence what feels out of your control   55

The challenge of being everything to everyone   70

Is a busy brain a clever brain? 91

Getting results easily and with less effort   106

PART 2 – YOUR COLLEAGUES AND CLIENTS   125

Working and living in balance   127

What to do when everyone demands innovation 193

Appearing competent to others 209

PART 3 – YOUR COMPANY 225

Leading with your brain switched on 227

A culture of more than psychological safety   243

Creating a culture of psychological safety yields an organizational environment in which employees can explore and express their full brain potential.   255

Managing people, managing brains   257

Preface

There are many brilliant people whose professional and personal lives are not living up to their expectations. Some feel cheated; perhaps they thought there was more to grown-up life. Others know they can do and be better and are prepared to work to achieve this. These people tend to have been on training courses, tried various tools and techniques and probably read several books claiming to have 'the answer'. While some of the things they've come across have been useful, they sense there is more.

This book doesn't pretend to offer you life's secret. Instead, it aims to offer you another lens through which to look: one which can unveil truths that people find very reassuring and encouraging. Typically, people respond with relief when they understand that there is a scientific underpinning to how they are feeling and behaving.

Over the past 20 years, our understanding of what drives our behaviours, and subsequently delivers our results, has increased massively, new research that deepens our knowledge comes in weekly. Most people are completely unaware of this incredibly useful information. I am frequently shocked during keynotes and workshops when significant foundational concepts about the brain come as new information to people.

Unfortunately, outdated research from 50 years ago is still being taught and tools that have been disproved to work are still being favoured. I see this in organization's people strategies every month.

This book will equip you with knowledge about how your biggest resource – your brain – works. This will help you work more effectively, productively and efficiently and give you what you need to perform at your optimum level more of the time. You are the expert in your life, your relationships and your organization. This is a guide to support you in becoming more adept at living and contributing in the way you want to.

Understanding your brain enables you to perform at a higher level in everything you do.

Acknowledgements

This book has only been possible thanks to the generous help I've received over the years. The insights I've been privileged to see have been thanks to the many individuals in organizations who have shared their experiences with me.

My deepest thanks and admiration go to the following people for the wisdom and energy they selflessly share with me. I am lucky to count them as my closest friends. Ben and Kate have tirelessly answered questions, shared their thoughts and been gently direct with their feedback all of which I hugely appreciate. Jessie has provided countless comical insights into human behaviour and always helped keep me grounded in good spirits. Rachel and Rob are such dedicated human beings and their wisdom, reflection and challenge are so enriching to my life and work.

Finally thanks to Stuart and Jessica, who make life so rich, playful and meaningful. I love exploring the world with you both and continuing to learn every day with you.

Introduction

Your brain is an amazing organ. It is key to you feeling that incredible sense of achievement when you help someone out when they're stuck, see a child succeed at something for the first time and even when you win a new contract. It is instrumental in you making decisions from whether to work late or go home and spend time with your family to whether to have that chocolate bar at lunch or go to the gym. Your brain is centrally involved in the important things in life, whether you are fulfilled, living a life on purpose, reaching your potential and being who you are made to be.

This book is all about making your brain work, about your relationship with yourself and then with others. You have an incredibly valuable resource in your head and most people don't have much idea how to work with it optimally. Embarking on the journey of improving how you use your brain is only the start. The much bigger questions come when you consider how you use your working brain and these are completely up to you. Deciding what you want to be efficient, effective and productive at are big life questions and ones that deserve mindful attention for the rest of our lives.

But is it even possible to 'make your brain work'? Isn't it doing its own thing? Aren't we programmed by our genes and fixed by our personalities?

This is a question I come back to almost every time I deliver a keynote talk or work deeply with an organization. What is really possible? What is practical? I believe that life is precious and that we should invest our time in worthwhile things. I avoid encouraging people to do things that don't have really strong evidence behind them.

I recall at an interview for a medical school I was asked what facility I would create if money were no object. I'd just read Francis

Crick's book *The Astonishing Hypothesis* and was passionate about what DNA profiling could unveil to us and how it could help people. Today I'm sure that wouldn't be the answer I'd give any more. This is because of the amazing new evidence emerging from the field of epigenetics.

You'll read compelling evidence in this book that you can indeed proactively 'make your brain work'. Does this mean that genetics no longer matters? Of course not. It is very important. But, the latest discoveries in epigenetics remind us that our environment is also critical. In fact some of my trusted colleagues would suggest that our environment accounts for up to 90 per cent of who we are. Whatever the figure – that is what we focus on. That which we can do something about.

My hope is always that my work helps you and those you care about. You are valuable, your ideas and efforts are appreciated, whether or not you always feel that way.

The fictitious characters that feature in this book – Kate, Jessie and Ben – all know they have untapped potential. They take responsibility for tapping into that and know that by understanding more about how they work they will be empowered to design and create more of the life that they want.

While the majority of the examples concentrate on business, the applications of the ways to make your brain work are widespread. For example, the things Kate learns as applied to the context of her work she can also apply to enrich her interactions with her children. Jessie would be able to use lessons she takes away to become more confident in her own skin. Ben has the opportunity to implement his new knowledge to improve his relationship with his new wife.

Neuroscience, the scientific study of the brain, underpins the content of this book. It is a wonderful discipline that has dramatically advanced our understanding of how we work over the last twenty years. There is a long road ahead of it still where many more insights will be discovered. As amazing as it is, it is only one of many areas of study that reveal truths. Learning about what it offers us and then referring to other subjects such as quantum physics or philosophy or theology can paint an even richer picture.

How to read this book

This book is designed for you if you are a busy professional who wants to get the most out of yourself. Each chapter has a range of different components to it, described below.

STORY TIME

Stories are easier for our brain to remember and help us understand concepts. They can enrich your experience of both the scientific underpinnings and the examples that each of the characters is going through.

EXPERIMENT TIME

Experiments are the backbone of neuroscience. They differentiate science from other disciplines by helping us to test theories to deepen our understanding. Reading bout key experiments gives you the power of knowledge to apply the results to your life.

FOR YOUR INFORMATION TIME

These sections are included to give you important information that will best help you understand what else is going on. They may be purely interesting or may have more necessity to them in expanding on neuroscience topics.

ORGANIZATIONAL INSIGHT TIME

This section is designed to give you a sneak peak behind the scenes of organizations we can learn from. They can help us wrestle with what might be best for our context.

MAKE YOUR BRAIN WORK (MYBW) TIPS

The end of each chapter finishes with tips that highlight some of the key points that will remind you of content from the chapter. These can be useful to refer back to and to work into an implementation plan.

You may choose to read the whole book at once, or more likely choose to go through it chapter at a time noting any insights you have about changes or applications to experiment with in your own life.

Meet the coach

Stuart is a trained coach, so understands how to get the most out of people and facilitates them being the best they can be. Having gone through a 'Neuroscience for Coaches' programme, he also understands the neuroscience that underpins how people work. When he works with professional people he plays many roles. Robert Dilts is a great researcher and thinker in the field of coaching, among others. In his excellent book *From Coach to Awakener* he describes the task of a coach as being to provide the 'necessary support and "guardianship" to help the client' successfully develop, grow and evolve at all levels of learning and change (Dilts, 2003). The levels Dilts identifies are: environmental, behavioural, capabilities, beliefs and values, identity and spiritual. In order to do this a coach may need to take on one of several different roles: guide, coach, teacher, mentor, sponsor and awakener. During most of Stuart's interactions with the three people in this book he acts in the role of guide, coach, teacher and mentor.

It is important for you to understand how your brain is working so you can optimize your performance. Ultimately this improves your quality of life and enables you to be happier in everything you are doing. We are naturally curious beings and like to understand how we work and how others work so this process and style of coaching often works well for professional people.

Meet the professionals

- Kate is a 54-year-old senior manager in a corporate position. She is divorced, has two grown-up children and is engaged to a lovely guy she's been seeing for three years. Her work, family and friends are everything to her.

- Jessie is a 32-year-old social entrepreneur who provides services to GP surgeries. After leaving a promising career as a doctor she has built this business up and now has a team of 20 staff. The business is growing rapidly and she is very passionate. She is single and enjoying living on her own for the first time after being with flatmates for years.

- Ben is a 26-year-old accountant, working in one of the Big 4 accountancy firms. He recently got married and tries to juggle the responsibilities of being a new husband with being a good employee and moving up the corporate ladder.

Part 1
YOU

Your productivity, efficiency and effectiveness are all under your control. This is both exciting and a big responsibility. Part 1 of *Make Your Brain Work* addresses the individual challenges professionals like Kate, Ben and Jessie experience and what is happening in their brains at that time. It empowers them to change their action as a result of what they learn. Ultimately the aim is to help them sort out the things that are most frustrating them and holding them back.

We explore ways you can see new opportunities and control your world from the inside out. Learn how to get a handle on your states to enable you to enjoy life and your relationships more. We look at a strategic approach to trying to be everything to everyone that brings the additional benefit of increasing your integrity. Dive into optimizing how you learn new things and equip yourself to make more of a difference during your lifetime. Also master the neuroscience of habits so you can enjoy the extra brain space and energy you have for challenging tasks. These are some of the components covered in Part 1 of *Make Your Brain Work*.

1
Can a marshmallow predict your success?

What happens when planning isn't enough and your sense of control is giving way to a feeling of being overwhelmed

Kate recalls Monday morning to her coach. Kate is qualified as a chartered surveyor and is now a senior manager in a global real estate services firm. Some people see her job as boring, but Kate knows that the people side of it keeps her on her toes. The day had started very well. She is in line for a promotion and is really excited about the upcoming week. It will be a chance to show her boss what she is really capable of. She had kept her diary after work clear so she'd have plenty of time to relax at the end of each day.

People had been noticing that she worked hard and she knew how important someone's reputation is in the corporate world. Everything was competitive. She always strived to maintain a cool outlook on things, not wanting people ever to see her flap. Keeping a tight lid on her emotions seemed to make her feel as if she could control what people thought of her.

When she arrived at work some of her colleagues were there already, perhaps she should have come in earlier she found herself thinking. Hurriedly, she sat down at her desk and turned on her computer. As the e-mails started pouring in she saw her previously tidy inbox become filled with questions, problems and appointment requests. She had been feeling calm but as she looked at the list she'd left from Friday that she needed to get done today she started to feel agitated. Her breathing became shallower and her stomach muscles felt as if they were tightening.

As she stared at the screen of e-mails, opening a few, seeing that they'd take too long to sort out and moving on to the next, the secretary she shared with a couple of colleagues in her team popped her head around the door and informed her of a last-minute team meeting this morning at 11.00 am. Briskly acknowledging this additional time stealer Kate decided that she needed to get on with something to get it out of the way. Suddenly she remembered she needed to write a letter to her boss about why she wanted this promotion. She picked one of the jobs from Friday's list and started doing the client work that was fairly easy, just to get something done.

The phone rang and her secretary wanted to put through a new enquiry with some big questions. Always wanting to seem open to new work, Kate took the call. She felt distracted as she looked at the clock knowing there were several big e-mail tasks that needed her attention, she wanted to be in control for the meeting at 11.00 am and there was still client work waiting from Friday. As the gentleman on the other end of the phone explained his situation she made a few notes while also checking a few more e-mails.

As she hung up she added another round of tasks to Friday's list and felt her head start to pound gently. As she finally completed the client work from Friday she moved on to today's post. As she looked at a familiar file it dawned on her that there was a potentially huge problem she hadn't considered relevant for this case at all, but did that mean she'd missed something. Once again she felt her breathing becoming constricted and it became a struggle to think clearly.

Situation

So what is the situation for Kate on this Monday morning? Stuart, her coach, identified with her the following things to focus on:

- She felt she had too many e-mails to deal with.
- Her 'to do' list was too long when combined with the other things needing to be done.
- She took a scatter-gun approach to the tasks.
- She felt a pressure to get started and tick tasks off.
- She felt worried about writing the letter well enough.
- She was trying to multitask.
- She had symptoms of being overwhelmed several times, including when seeing something new to her.

This chapter helps you to understand how your brain gets overwhelmed and how instead to be in control in a way that enables you to come up with great creative ideas and solutions, eliminate interruptions and produce better quality work. The bonus is increased efficiency and effectiveness.

Possibilities

Stuart starts by looking at Kate's feeling of being overwhelmed. Different people get overwhelmed by different things and in different situations. Some people may need to be training for a marathon, managing several accounts, be a parent to a new baby and then have to deal with their car breaking down in order to start feeling overwhelmed. For other people being asked if they'd like a BLT or tuna melt sandwich could be the thing that adds just that little bit too much to their already full mind and makes them feel overwhelmed.

We can almost think of the brain as having a strategy for becoming overwhelmed. Rather than taking the approach of suggesting to

Kate that she just tries to keep calm and take things one at a time, which tends to be next to useless, we're going to look at what is actually happening when a person gets overwhelmed and, as a result, what that person could do differently. General high-level advice is one thing, and occasionally may be effective, but understanding what is actually going on, for everyone, from a neuroscience perspective means that you can draw out a plan that you have designed definitely to work for you.

Let's jump right into meeting the main player when we're working with becoming overwhelmed. The central thing for us to be aware of is actually an area of the brain. It is called the prefrontal cortex (PFC).

FACTS ABOUT THE PREFRONTAL CORTEX

It is right at the front of the brain, just behind your forehead. Its role is classically described as like that of a CEO (if you are business minded) or a conductor (if you are musically minded). In short, it is the boss, it is responsible for your 'executive' functions, which means your ability to think, choose, plan etc.

Over the years it has developed considerably, and recent studies indicate that meditation increases its size further still.

It is hugely energy hungry, but gets drained quickly. Stress impairs its ability to use energy.

Structurally, it is part of your frontal lobe, which is the area at the front most part of your brain.

Prioritization

Stuart feels it is important that Kate looks a little more closely at her prioritization. He asks her what she thinks about it, whether she ever does it, how she thinks it works. Prioritization is one of those things that everyone knows they should do. They know intellectually that it's supposed to be a good idea, but few people do it effectively.

Kate's lack of understanding about how prioritization actually works means that when she is doing it, she isn't doing it in a way that will help her. By feeling pressured to start ticking things off her list she tries to jump right in to doing tasks, but the downside to this is she starts using valuable prefrontal cortex energy. She is tiring out her CEO (the PFC part of her brain). Her CEO is actually best equipped to do the prioritizing. So by exhausting it, if she ever gives it the chance to do the prioritizing it will be too tired and do a very bad job.

Prioritization takes a lot of energy so it is best done with a strong focus. So how do we focus? (And 'I just focus' doesn't cut it as an answer here). Anytime you are concentrating, learning with great intent or seriously paying attention to someone or thing then your frontal lobe, of which your PFC is a part, is keeping a tight rein on your brain to stop it wandering off to other activities. The frontal lobe acts a bit like a bouncer – fending off any unwanted signals from the body, any emotions, and any environmental senses. The bouncer makes any sensory or motor information quiet, meaning that we tend to find ourselves being quite still. While the bouncer is outside fending off unwanted visitors, inside we are in a form of trance.

Picturing your goal

Stepping away from the analogy, let's look at the actual benefit of prioritizing. When you are being optimally efficient and effective you will experience a state that feels similar to the state of 'flow'. (The actual state of flow has several components to it, which may or may not be additionally present). What does this feel like? You feel still, things seem quiet and you don't have any strong emotions. In fact the thing you are focusing on becomes more real to you than anything else. Time passes without you realizing. Scientists call this whole process lowering the signal-to-noise ratio. You can imagine how helpful this state could be when you want to get a big important task done, like the important letter Kate had to write, which was really going to stretch her prefrontal cortex. It would mean she would feel relaxed, focused and could proactively work through the task without constantly having fleeting thoughts of other things she needed to do jump into her head. The end resulting letter would reflect her best work.

Neuroscientists believe that being able to see and hold in your mind a picture of what you want to do makes it possible to then do it. This is directly linked to the emotional and motivational processing of this intention. When prioritization takes place it means you then have the programming telling you when and under what circumstances you will do the task.

Another component of prioritization can involve the sacrificing of immediate gratification for delayed gratification. You need to be able to place value judgements on your outcomes in order to choose which to work on first, and to place enough value on something to do it at all. Have you ever found yourself wishing you were the type of person who went to the gym regularly, or meditated or ate more vegetables or remembered people's birthdays or anything else? But you found yourself week after week, month after month, even year after year not doing it? This again comes down to your frontal lobe to sort out. It is not that you don't really want the result that going to the gym will bring; you absolutely want a healthy, fit, toned body; it's more complex than that. It may be that your brain has equated gym attendance with a poor relationship (because most people you know who regularly go to the gym are single or in unhappy relationships). It may be that when you do go to the gym you work out so hard that you are in agony every time you laugh or walk for days afterwards. It may even just be that you want other things (like a bottle of wine and rest in front of the television every evening more than you want the healthy, fit, toned body).

CAN A MARSHMALLOW PREDICT YOUR SUCCESS?

The Stanford Marshmallow Experiment leads us to believe it can. In 1972 a psychologist called Walter Mischel conducted an experiment with around 600 pre-school children. Individually the children were led into a plain room, devoid of any distractions, that had their choice of treat (Oreo cookie, marshmallow or pretzel) on a table by a chair. They were told that they could eat the treat and ring the bell at any time (signifying they had eaten it) or wait until the researcher returned, at which point they would be rewarded with a second of their chosen treat. Watching the children wait is quite an experience. Some cover their eyes with their hands, others tug their pigtails, some stroked the treat, as if they were caring for it, and some even start kicking the desk, while some turn away from the temptation.

A third of the children could delay their gratification long enough to get a second treat. This is interesting, but what is even more interesting is the follow up study, which showed that the child who could wait 15 minutes had an SAT score that was, on average, 210 points higher than that of the child who could wait only 30 seconds. To dive deeper into this research check out *The Marshmallow Test* by Walter Mischel.

Starting with the end in mind

Our ability to delay gratification or to exercise self-control has a correlation with achievement. Our anterior prefrontal cortex (a part of that familiar CEO again) is involved in abstract problem solving and keeping track of goals. Yale University researchers recently scanned 103 people and found that delaying gratification involves the ability to imagine a future event clearly. One researcher, Jeremy Gray, said you need 'a sort of far-sightedness'.

If Kate is constantly feeling the need for immediate gratification to feel good about getting 'on with something to get it out of the way' to 'just get it done' then the bigger tasks just aren't going to get done. Calling them bigger tasks isn't even fair, because even if they start as bigger tasks part of the process should be to break them down into manageable parts.

Classic productivity teachings steer us to beginning things with an awareness of what the end looks like and the marshmallow experiment demonstrates the huge importance of this. The advances in technology mean that we can see that the anterior prefrontal cortex is vital for being able to do this.

Thousands of experiments including the classic brain scan, functional magnetic resonance imaging (fMRI), allow us to see this area of the brain lighting up when a person is performing cognitive functions. We also learn from experiments with people who have brain damage, such as Phineas Gage who became unable to make good decisions after sustaining an injury to this part of his brain. Later in this chapter we will look at what Kate can do to keep her prefrontal cortex working optimally.

Prefrontal cortex

So what is the motivation for ensuring that our PFC is in optimal working condition? When our PFC is not working optimally we find ourselves:

- feeling lazy;
- feeling lethargic;
- uninspired;
- easily distracted;
- being poor at completing things;
- fixing attention on repetitive negative thoughts;
- being disorganized;
- being forgetful;
- feeling overly emotional.

On the other hand, when Kate's PFC is working in tip top condition she can look forward to:

- intentional awareness;
- a long attention span;
- being able to contemplate possibilities;
- being able to plan;
- being able to stick to the plan;
- focusing easily.

When the PFC is overworked it doesn't function well at all. All the symptoms mentioned previously can occur and this makes it very difficult to be effective, let alone efficient, which is often very frustrating to the people trying to get through the day. During this time it's quite common to feel as if something is wrong and perhaps to fall back on old habits. This is a form of survival mechanism. When people find themselves micromanaging, controlling or punishing there is often a brain deficit involved. This doesn't always mean something is overactive; an underactive PFC causes problems too. Dopamine is a neurotransmitter that has many functions for the brain including reward, motivation, working memory and attention. When the brain has a lowered ability to use or to access dopamine it means that other brain areas aren't being quietened down to enable the brain to focus on one thing at a time. This makes life difficult for us because we struggle to focus, so become less efficient.

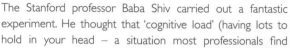

CAKE OR FRUIT SALAD

The Stanford professor Baba Shiv carried out a fantastic experiment. He thought that 'cognitive load' (having lots to hold in your head – a situation most professionals find themselves in) might influence self-control. He gave half his volunteers a two-digit number to remember (representing a low cognitive load) and gave the other half a seven-digit number (a high load). The volunteers were then told to walk to another room in the building and in so doing pass a table where they had to choose between chocolate cake or fruit salad. Of the people with the high load, 59 per cent opted for cake whereas only 37 per cent of the people with the low load did.

Shiv speculates that remembering seven numbers required cognitive resources that had to come from somewhere, and in this case were taken from our ability to control our urges! Anatomically this is plausible because working memory (where we 'store' the seven or two numbers) and self-control are both located in our prefrontal cortex. The neurons (brain cells) that would normally be helping us make healthy food choices were otherwise engaged in remembering seven numbers. In those instances we have to rely on our more impulsive emotions, such as 'mmmm yummy – chocolate cake please'. Stuart is aware that prioritization is this energy intensive process, and ideally would be best done in the morning...but in time Kate may decide to trial this herself, or he might suggest it.

Practical prioritization

The most important thing Kate takes away from this is that she needs to get clear on the things that she wants to do before just jumping into doing things. Prioritization needs to be her priority! She decides to set aside 10 minutes at the start of the day to look properly at what order she wants to do things in and to visualize the end result of each of them. Stuart is aware that prioritization is this energy intensive process, and ideally would be best done in the morning...but in time Kate may decide to trial this herself, or he might suggest it.

Since Kate is serious about this she ensures that she puts the 10 minutes in her diary. There are many ways to prioritize. Kate naturally finds that writing out a big list of everything that is in her head is useful. Then she goes through the list and puts a dot by anything that definitely needs to happen today. She briefly closes her eyes and asks herself what she will see, hear and feel when she has completed each of the dotted tasks – this creates a strong connection neurologically with the task she wants to achieve. She can also at this point get a feel for how long it will take her to complete each task. Once she has all these components she can instinctively number the tasks in order of how they will fit together best. Anything that isn't vital for today or doesn't fit into today gets transferred to a new list for tomorrow or another day.

Knowing that you may not get the prioritization spot on at first is a comfort and Stuart recommends that you work on two-week trial programmes. So you choose a strategy, trial it as you would a scientific experiment, not varying anything but being observant about how it works for you. Then at the end of the two weeks evaluate what went well and what could be improved upon. Decide on new parameters and begin another two-week trial.

New information

When Kate gets a call from another client she starts to feel uneasy again. She is hearing new information; they are mentioning things that haven't come up before. This is very common. Imagine a situation where a boss or a colleague tells you something that was new to you. Your interest levels skyrocket. You pay attention. There is a very important chemical at work inside you at this time, which we'll come back to later. For now, let's look at the area of the brain that is getting excited.

THE NOVELTY FACTOR

Novelty sends the prefrontal cortex crazy, if it is just ticking along something new really wakes it up. A neurologist called Marcus Raichle did several experiments to find out what was going on. He gave individuals a task that required them to say an appropriate verb to a visually presented noun. Initially when they were presented with the task their blood flow to their frontal lobe reached its highest level – indicating lots of activity. As they continued to do the task the frontal lobe chilled out and their involvement (based on blood flow) pretty much disappeared. A new task was then given to them, similar to the first, but with differences. The blood flow increased again, but not to the same level as before.

Summary: the frontal lobe is involved when things are novel. You will remember things more easily when they are novel. Your engagement with novel things is higher.

When things are new your frontal lobe is being stretched so you need to keep everything else as simple and streamlined as possible to reduce the pressure on them (otherwise you'll feel overwhelmed and struggle to process anything). In Kate's mind she is merging everything together. This doesn't help her brain process the new information. The best thing when confronted with new data (in whatever form) is to separate it as something new. Once this is done you can then break down what this new information is and link it to things that are familiar to you. If you don't want to be using up your thinking power on something new then looking for ways to link it to something familiar is great because it does not then create the same level of frontal lobe activity.

For example, if you are told that you need to start 'tweeting' your frontal lobe will go into overdrive if you've never done it before. This in itself is fine, but you would start to feel overwhelmed if you were also thinking about all the reports you need to write, the people you need to schedule meetings with and the fact that you promised to take a holiday in the next month. The best thing to do in this situation is to recognize that 'tweeting' is new for you, but you're great at LinkedIn, which is similar. When people ask 'what's that like?' they are looking for something familiar they can connect with. Unconsciously Kate will now be considering that she learnt to get good at LinkedIn, which means that she can learn to be good at Twitter.

Multi or mono

The dreaded multitasking – it is at epidemic levels. Kate is quite normal to try to do it at work, at home, even in bed. The famous Blackberry (and now the iPhone) enticed people to up their game in the multitasking arena. It wasn't enough to reply to e-mails while in the office and on the phone, something Kate regularly diced with. Now Kate regularly is on the phone to her mother, cooking dinner, checking Facebook on her laptop and checking work e-mails on her iPhone. All the while she is keeping a 'relaxing' television pro- gramme on in the background to 'help her switch off'. If it didn't have such important consequences it would be funny!

The prefrontal cortex, like a good conductor, likes to focus on one thing at a time. Conductors never play a bar of one piece of music then move on to play eight bars of another piece. Every time a con- ductor decides to play another piece all the musicians would have to put their instruments down, find the other piece of music from their folder, pick up their instruments again then start playing. So in moving from one piece of music to another the amount of time wasted would be huge and it takes a lot of energy to keep switching. Equally, if the conductor thought he could save time by practising two pieces at once, you'd get a big jumble. Bars from each piece would have to be missed out. Things would be confusing. Neither piece would be played well, and it would be hugely tiring for the musicians. This is exactly what happens when Kate is trying to mul- titask. She misses bits of what the client is saying. She can only scan the e-mail, missing bits of that when she is tuning into the client on the phone. Afterwards she has expended a huge amount of valuable energy and feels drained.

MULTI-TASKING PERHAPS THE NEW ENEMY OF PRODUCTIVITY

A professor of mathematical psychology called David Meyer took a group of young adults to test what happens when people are switching between things quickly. The experiment involved the participants working out mathematical problems and identifying shapes. When they had to switch between the tasks their accuracy and their speed decreased compared to when they could perform one and then the other. In some cases multitasking added 50 per cent to the time required. Imagine a person working a 12-hour day and achieving the same amount of results as a person working 8 hours, but with more mistakes and less elegance. One comment was: 'Not only the speed of performance, the accuracy of performance, but what I call the fluency of performance, the gracefulness of their performance, was negatively influenced by the overload of multitasking.'

There is no conclusion other than to get excited about mono-tasking – it'll be the new multitasking – only more efficient and effective and subsequently more productive.

MONOTASKING MEETINGS

I am frequently being told by individuals in organizations that they feel pressure to keep on top of e-mails during meetings.

There are so many meetings and so many e-mails. After a day filled with meetings one can return to hundreds of e-mails. This is simply crazy. Unless the plan really is to work 18 hour days. Which is also crazy.

So this problem isn't going to be solved only through more intentional management strategies. It needs some overall cultural and behavioural shifts within organizations. Fewer e-mails need to be sent. Fewer meetings need to happen. Often, less work needs to happen. This may not be what some companies want to hear – but would you want someone you cared about to work so hard they didn't see the people that were important to them in life, or when they did they were exhausted and irritable? Would you tolerate someone increasing their likelihood of a heart attack? We have to remember what is important.

An organization that just keeps wanting people to work harder and longer isn't honouring human beings brains or bodies. Should these kinds of companies exist?

Your plastic brain

Michael Merzenich, who you will meet again later, is probably the world's leading researcher on the best thing since sliced bread: neuroplasticity. We will cover this in more depth in Chapter 3. However, for now it will benefit Kate to know that her brain can, and indeed does, change.

As a pianist Kate is now an elegant and graceful musician. It wasn't always like this though. When she was learning as a child she used her whole upper body, her face was screwed up and her elbow and shoulder seemed to jar with every note. At that time a massive number of neurons (brain cells) were required for her to play each note. Now, she only requires the specialized ones that are very good at it. Obviously this is vastly more efficient.

MONKEY BUSINESS

Merzenich is famous for many experiments with monkeys. In one he trained a monkey to touch a spinning disk with a certain amount of pressure for a certain amount of time. He was then rewarded with a banana pellet reward (the monkey, not Merzenich). The monkey's brain was mapped before and after the experiments. What happened has huge implications. The overall area of that particular map in the monkey's brain got bigger. This makes sense as more brain resources are being dedicated to the more frequently carried out tasks. The individual neuron's receptive fields got smaller – more accurate – and only fired when small corresponding parts of its fingertip touched the disk. So there were more accurate neurons available to do this task.

Here's where things get really fascinating. Merzenich found that as these trained neurons got more efficient they processed faster. This means that our speed of thought is plastic. Through deliberate, focused repetition our neurons are being trained to fire more quickly. They also don't need to rest for as long between actions. Imagine how much more powerful and effective you would be if you could think quicker. It doesn't even stop there, the faster communications are also clearer, so more likely to fire in sync with other fast communications ultimately making more powerful networks. More powerful networks or messages make it more likely we'll remember something.

The final piece in this research extravaganza is that these brilliant changes, which are available to you, only stick long term if you pay conscious attention to the tasks as you do them. So if Kate wants the benefit of faster thinking capacity and the ability to recall things easily in the future then she needs to pay conscious attention to one thing at a time. This is relatively easy to do, with practice. You'll have experienced times when you are really present, perhaps at an important meeting, or while listening intently to a friend, or watching an important match. Your focus is on what you are doing. You are actively present. In contrast, you'll be able to remember a time when you think you are reading something, but really none of it is going in because you are thinking about what you're going to have for lunch, or do at the weekend, or another important paper you need to read. This is the difference between paying conscious attention and unconsciously just moving through things.

Reassuringly flexible

Kate starts to wonder how we can even switch our attention between things at all and Stuart takes this opportunity to build up Kate's faith in her brain as being designed to serve her best, so long as she chooses to use it in the way it is designed.

The brain has the ability to maintain a set of mental activities, such as reading an e-mail, in a stable way. It can do that for a period of time and do it very well. Then it can make a quick switch to another set of activities, such as listening to a colleague, or talking on the phone. These will be maintained in an equally stable way. This is called 'dynamic bystability'. Switching backwards and forwards quickly, as happens during multi-tasking, still attracts a big energy cost. The way the dynamic bystability works is still being investigated but we believe it has to do with two forms of dopamine (a neurotransmitter) receptors. One of which has a stabilizing effect and the other has a destabilizing or updating effect.

THE E-MAIL BRAIN DRAIN

Organizations are slowly switching on to this massive time drainer.

Wednesday walkabouts – don't send any e-mails unless absolutely necessary

Weekend e-mail ban – no e-mails can be sent over the weekend (can result in barrage Monday morning)

Enriching your potential

Stuart shares a particular experiment that he hopes will inspire Kate to boost her brain potential so she can cope with more things before becoming overwhelmed. Knowing that Kate really wants this promotion helps guide her coach to share the most appropriate stories and research that will best serve her. Sometimes Kate feels that because there is so much to do she shouldn't go and see her friends, or do the things she enjoys such as go to the theatre or go for massages. Stuart wants to show her that this isn't the case and this experiment is the first ingredient in showing her.

LIVING THE LIFE OF RAT LUXURY

Back in the 1970s a neuroscientist by the name of Bill Greenough did some experiments with rats and their living accommodation. One poor group of rats drew the short straw and ended up living alone with nothing to do. The other group were bestowed comparatively plush surroundings. They had exercise wheels, ladders to climb, and other rats to talk to. Greenough called it 'the rat equivalent of Disneyland'. These lucky rats soon became noticeably more physically and socially active, as far as laboratory rats can.

Things became really interesting when their brains were later examined. The 'enriched' environment rats had 25 per cent more synapses (connections between a neuron and another cell) per neuron than their poor relatives. These additional synapses meant the rats were cleverer and quicker to find their way through mazes and were able to learn landmarks faster.

To reduce the problem of becoming overwhelmed it's important that we look at both the short-term strategy for getting rid of the problems being faced but also the long-term strategy that will reduce the frequency of the problems arising. By creating your world as your very own Disneyland, whatever that means to you, you are going to upgrade your brain, making it easier and quicker for you to work things out in the future.

Action

Kate decides to make some changes based on everything she learnt and the things she reflected on during the coaching session. She makes a list of all the things she wants to change eventually, and then picks just one a week to start implementing. If she finds she struggles with it one week she plans to continue focusing on that change for a second week before adding another thing into the mix. She finds a list entitled 'to do' not quite as inspiring or uplifting as 'action' so names hers accordingly.

Kate's action list

- Write the letter for the promotion first thing in the morning, with e-mails and phone turned off.
- Decide on times of the day to tackle e-mails (10.00–11.00 am, 2.00–2.30 pm, 5.00–5.30 pm).
- If there isn't enough time to tackle all the e-mails in the slot allocated schedule another time to look at the ones left over.
- Choose and schedule the big priorities for the week on Sunday night or Monday morning.
- Schedule 10 minutes each evening to prioritize the following day.
- Check that plan each morning, make sure you know your priorities and get them done first.
- Practise doing one thing at a time and paying conscious attention to it.
- Go for a massage this week.
- Trial everything for two weeks then re-evaluate.

MYBW top tips to prevent being overwhelmed

- Turn off the e-mail function of your mobile phone in the evening so your brain has down time in the morning before you start work.

- Prioritize the big weekly tasks first then smaller tasks on a daily basis (experiment with the night before or first thing in the morning to see which works best for you).
- Turn on e-mails only for certain windows of the day.
- Mono-task for short- and long-term benefits.
- Decide that you are in charge of your time and acknowledge that you are choosing how to spend your time.
- Become your own best detective, identifying how you get overwhelmed and how you avoid it.

MYBW top benefits for mastering your feelings of being overwhelmed

- Come up with great ideas during quiet time when you're not being interrupted by e-mails or calls.
- Save time that previously you'd have spent thinking about what to do next every time you start a new task during the day. Also be more focused when you start a new task because your unconscious mind has been working on it for you overnight.
- Increase your efficiency and effectiveness by eliminating interruptions.
- Increase your perception of autonomy, by enabling you to be more in control of your time, can lead to more productive action and better quality work.

2

Is your hippopotamus under attack?

What is really going on when everything seems to be going wrong and your stress levels feel as if they're soaring?

Jessie prefers to meet her coach Stuart in person and has been doing so since her entrepreneurial business, which provides services to GP surgeries so they can better serve their patients, was 6 months old, almost 18 months ago now. She finds it really helpful to have someone work through her approach to things, how she is thinking and how she can be more effective. Each week she tends to work on something different and this week she knows exactly what she wants to talk to him about.

Last Friday Jessie had received a very confused phone call from one of her GP surgeries. Basically there was a mix up – fitness coaches had turned up to a group of ballroom dancers and the dancing coaches had turned up to the keep fit group. Apparently the faces of the elderly people dressed in their finest ballroom outfits when the high energy coaches in their bright lycra jumped out of the van with their dumbbells was a picture. Similarly, when the other group, dressed in their baggy T-shirts and joggers, were confronted with the well-dressed dance instructor couple there were raised eyebrows to say the least.

Although they could eventually see the funny side, the fact is that this is just one example of things starting to come apart at the seams. Jessie is missing meetings, forgetting what she has or hasn't done and she is scared that a big mistake is just around the corner. When her PA made a jokey remark about the mix up Jessie almost bit her head off. All her old feelings of not being good enough came racing back. Later, when she was speaking to another client about their needs and the subject of the fitness coaches came up again she felt all those feelings of inadequacy rush back to her again.

Everything just seems harder at the moment, she has had a cold for almost two weeks and she often sleeps weirdly and has a sore neck and shoulders all day. She is finding it difficult to concentrate on things and to make sense of what people tell her. It's as if her thoughts are working at half speed. She just feels like going home, getting into her pyjamas and watching Netflix. Or even better, just escaping somewhere warm for a nice holiday, but that just isn't possible.

Situation

Stuart summarized the key points Jessie had mentioned:

- a feeling of uneasiness she couldn't put her finger on;
- feeling tense;
- her legs feeling heavy and neck and shoulders feeling sore;
- taking longer than normal to get over a cold;
- making a basic mistake;
- getting defensive with her PA;
- old 'stuff' coming up;
- 'pulling herself together';
- thoughts working at half speed;

- all the negative emotions coming up again when talking about the fitness coaches;
- wishing she could go home and watch *Netflix*.

This chapter is about learning to reduce your stress levels on a fundamental rather than superficial level to see new opportunities and avoid the negative spiral. The bonus is being able to control your world from the inside out. We need to be aware that stress is killing people. It is also crippling lives through physical and mental illness. The importance of this epidemic cannot be overstated.

Stress: the classic culprit

The biggest contributing factor that Stuart wants Jessie to work on is that her normal level of stress is too high. There are several types of stress that can be caused by many different things. We will focus on emotional and psychological stress. Interestingly, the situation that is emotionally stressful for one person may be invigorating for another. What is fun for you may be stressful for your colleague. This is because we all process things differently. We attach different meaning to things based on what we have experienced before.

When Jessie finds out about her error in sending the wrong teams to the GP surgeries she processes this as a bad thing. Another person may have just seen the funny side. To Jessie it means failure, doing a bad job and upsetting people. By attaching these meanings it is natural for her body to release chemicals into her blood stream. On a basic level she is feeling threatened. In caveman times the threat may have been a tiger or a bear, today it is often status or identity that is threatened. The effect of the chemicals surging around her body are to increase her blood glucose levels (giving her more energy), suppress her immune system (unnecessary at a threatening time) and increase metabolism of fat, protein and carbohydrates. The chemicals do a good job in helping the body deal with stress. They are the good guys; they work to restore balance in the body. However, when they stick around too long, as is the case of chronic stress, they start to do damage.

Good intentions

Jessie is feeling stressed because she has an 'allostatic load'. This is what happens when you are chronically stressed. As we've just learnt, when our bodies get out of balance for some reason, mechanisms kick into play to try to bring them back into balance. Allostasis is the body's adaptive response to stress. With just a little stress our memory tends to increase, but with a lot or with chronic stress our memory decreases. Having a high allostatic load also affects your mood, emotions and behaviours. Your outlook also shifts from a long-term outlook to a short-term one. These are all major things that are affecting Jessie's everyday life experience.

When suffering from chronic stress people tend to be reacting in a less than helpful way to their environment. The bottom line is that stress impedes learning, relating and achieving. It has been said that we suffer from chronic stress a lot. Perhaps you can remember a time when you felt under a lot of pressure and your productivity suffered as a result. Or maybe you were worried about a situation that you couldn't see a solution to. Maybe you can see symptoms of chronic stress in people you work with. Recognizing it is the first step.

The following symptoms are indicators of chronic stress:

- difficulty in concentrating;
- short temper;
- feeling nervous or worried;
- struggling with decision making;
- difficulty sleeping or insomnia;
- depression;
- lack of appetite;
- persistent ill health;
- muscle tension;
- having an upset stomach.

A person experiencing any of these for longer than a few days needs to take action to get things back into a healthy balance.

In healthy people, when they encounter a stressor (something that stresses them) their normal chemical balance gets disrupted. The body starts the stress response and this re-establishes the normal homeostatic balance. Stressors are different for different people. Take for example Jessie's colleague, Judith – hardly anything seems to affect her. There can be suppliers shouting at her and clients being rude to her and she still seems unaffected. For Jessie though it seems as if even the little things make her feel stressed. This is what happens when people are chronically stressed. They process things differently and are effectively on high alert most of the time. This means that their bodies aren't getting the time or resources to regenerate.

IMMUNE SYSTEM SCIENCE

Research shows that chronic stress leads to the production of a certain protein (called calcitonin gene-related peptide or CGRP). This protein gets up to mischief with our immune cells. A major set of immune cells, called the Langerhans cells, have the job of capturing infectious agents and delivering them to the infection busting cells, called lymphocytes. What the CGRP does though is to coat the Langerhans cells, which renders them unable to capture the infections. The result is that people are more susceptible to infection.

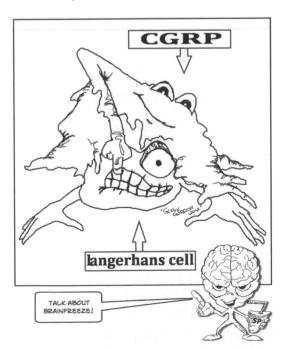

Some prominent scientists assert that almost every major illness that people acquire has been linked to chronic stress.

Of course when people get sick they miss work, which makes them stressed, which starts the cycle again. This is why it is so important to address chronic stress.

MODELLING MONDELEZ

A great recent privilege for the Synaptic Potential team has been working with Mondelez. This is the snack company, who kindly let me into the staff Cadbury's shop! They recognize that their people go through times of greater intensity. These may trigger more stress in some people. The organization takes seriously the health and wellbeing, as well as the performance, of their people.

Mondelez already had a great online University when we met them. It was incredibly impressive in our opinion. Yet they recognized that our digital content, underpinned by scientific research, would enhance their people's ability to deal with stress. They quickly understood the difference between dealing with stress in the moment and building resilience over time. (Something many organizations miss when lumping it all together).

Putting people through a half day workshop on stress management may do some good – but if you want real, sustainable results then a more thorough approach is required. It doesn't have to be expensive – it does have to honour how people work.

You can read more about how this was structured in Chapter 10 on resilience.

Predictability and stress

Stuart introduces a new key concept that could help Jessie reduce her stress levels. Our modern brains want security; they love plans, agendas, to know what is happening. Security in predictability is great for us. However, many jobs now frequently plunge people into insecurity. People may not have clear boundaries for their role, or know what is expected from them week to week. They may not understand how to do a great job in their boss's eyes or what their colleagues really think of them. New initiatives can be sprung on people without proper introduction or framing, causing people to feel unsettled. It doesn't have to be this way though; there are ways of introducing and communicating new things that could trigger a stress response so that they are avoided. Often leadership and management can play a big part in this in many types of jobs.

For entrepreneurs predictability needs to be managed internally, normally by the entrepreneurs themselves. Jessie can't look to a boss to give her security, because she's the owner of her business; she needs to create it herself. The entrepreneur's world tends to be a flexible one, where change and adaptation are strengths to be admired. These traits still need respecting and so the predictability needs to come from somewhere else. Stuart asks Jessie to make some suggestions:

- Implement a system for the accounts always to go to the accountant by a certain day of the month and arrange always to receive back the summary sheet on a certain day. This would make Jessie feel on top of the finances.

- Dedicate one day a month to external relations for members of the whole team to brainstorm press releases, Twitter feeds, blogs and anyone they want to connect with for the next month. This would ensure that it gets done and so be a weight off Jessie's mind.

- Have bi-monthly meetings with each member of the team to coach them and help them create their role to be more of what they excel at. It is hoped that this would reduce any sudden revelations that things have gone too far in the wrong direction.

Jessie starts to understand that in the past when she was feeling confused her brain was feeling threatened and this was compounding her stress. As she started to feel guilty about what had happened her amygdala (a part of the 'emotional brain') responded to this emotion. Guilt, like shame, envy, jealousy, sadness, despair and hate, is one of the more subtle social emotions. These are dealt with by the insulae (another part of the 'emotional brain'). A great thought leader in the field of emotions, Antonio Damasio, proposed that the insulae links body sensations to emotions. Some of Eisenberger's research at UCLA supports this. When Jessie is feeling guilty her neural pain circuits will be activated, causing her to feel the pain of guilt physically.

When Jessie doesn't feel secure she can feel a physical sense of discomfort. When she feels guilty about making a mistake she can literally feel pain in her body. These concepts are new and foreign to many people but are backed up by cutting-edge science. Now Jessie knows that to avoid that pain she can put things in place to give her some security and predictability. She can also still retain the times of creative flexibility where some of her best ideas come to her. In fact, she feels confident that these insights will occur more often now that she will be less stressed.

Perhaps pounding the pavements is the answer

So should Jessie start exercising to further alleviate her stress? This is something Jessie has heard before and most people would say 'absolutely' believing that exercise will help to reduce stress. This is one of the dangers of the generalizations of science that reaches people. Exercise absolutely **can** be good for decreasing stress levels, but it can also have other effects, as the study below shows us.

EXERCISE COULD MAKE STRESS WORSE

In the 1990s a study was conducted at Yale University. Two groups of actors had to get themselves into different emotional states. The first group made themselves angry by imagining frustrating and disturbing situations. The second group kept themselves calm and peaceful. Each group had their heart rate, blood pressure and respiration monitored.

Each group was then asked to do various forms of light exercise, such as climbing stairs. The angry group's physiological measurements were less healthy. The other group, the calm actors, enjoyed the physical benefits of the exercise. Most people believe that exercise is good for reducing stress, however your state of mind when doing that exercise is proving to be critical. This isn't to say that in an angry state you wouldn't still potentially feel a release when thrashing it out on a squash court, but physiologically it could be better for you to adjust your state of mind first.

So what states of mind would be good for exercise? Here are some classic swaps a person could make.

Old thought	New thought
Feeling irritated with a client for giving a new big piece of work to another company	Consider calmly the things you will work on practically to make your proposal more desirable for next time
Being angry at a boss for not promoting you	Focus on all the things you are grateful for and appreciate at your current place of work (either to find these somewhere else or just to refocus)
Thinking your colleagues are lazy and stupid and you're fed up with having to carry them	Consider who you would need to be to bring out the best in your colleagues, and what that would mean about your abilities
Thinking that you're failing	Objectively list the successes you've had over the last week then decide on three new actions you want to take to lead to more success

Changing your initial state of mind is a great first port of call for many reasons. It opens up new possibilities, improves your ability to think, enables your body to release more helpful chemicals into your blood stream and has positive effects on other people. Some people have evolved strategies that make this easy to do, while other people really struggle with it. It is possible to improve your ability to change your thoughts and change your states. We'll look at how this can be done at various points. Here are some classic practical techniques. They look straight forward but don't underestimate their power. If you could expertly do each of the following then your life and your relationships would be greatly enhanced.

- **Refocus** – shift your focus from where it currently is to something that is completely different or just a different area of the same thing. If a client phones you with a long list of problems, refocusing may involve zoning in on any positive things they did say or on another client's feedback, which was positive. Normally this is best done after addressing anything that you needed to action based on negative client feedback. Refocusing often involves looking at a bigger or smaller section of the situation.

- **Reframe** – changing the frame around something often changes how it looks and feels. With photographs of people you know you'd be reframing by thinking about the good times you once shared with them. In business your ability to change your frame and help others change theirs is very valuable.

- **Revaluate** – this involves reconsidering the meaning you've attached to it originally. This is often the most difficult for people to do because on some level they resist the change (and the idea that their initial evaluation wasn't correct). Taking the photograph example, it would be like saying that the people are still in your life in the form of memories and the impact they've had on you.

Practise your ability to see things differently regularly. Offering alternatives to what everyone else sees will help you stand out and add value in many different situations. When everyone is calling a colleague a gossip, perhaps you change their focus by asking why they think the colleague feels the need constantly to talk about people behind their backs. Questions are powerful; use them generously.

Negative connotations

When Jessie has to go and see a different client who talks about fitness coaches it just brings back all the negative feelings she had tried hard to suppress. This is a form of anchoring, the mechanisms of which we'll talk about more in Chapter 3 when we meet Ben.

DEADLY ASSOCIATIONS

A group of Russian researchers performed an experiment on rats that shows us just how powerful associations can be. They set the rats up to ingest an immune-suppressing drug that was flavoured with saccharine (an artificial sweetener). A side effect of the drug was to make the rats feel sick. They were given this combination many times. Then the researchers stopped giving them the drug, and just gave them the saccharine. Here's where it gets interesting; the rats still got sick. They had been conditioned to associate the saccharine taste with the physical symptom. Many of the rats actually died. Their thoughts weakened their immune system so they were defenceless against their environment.

Stuart is keen for Jessie to unhook the connection between feeling bad and fitness coaches. Every time Jessie feels bad about something it has an effect on her body and mind. The first time we feel bad there is often a positive intention, perhaps something for us to learn, some action we need to take. After that the feeling bad rarely serves a purpose, which means we should work on getting rid of it because it has negative side effects. This is just one example. Every day you are experiencing things and associated feelings with the things and most of the time this occurs unconsciously.

Checking out of stress and into normality

Jessie summarized her feelings about her day by saying she wished she could just go home, put on her pyjamas and watch netflix. This is her craving familiarity. It is really common; perhaps you know what your friends or colleagues do when they feel in need of familiarity. Let's look at what is happening inside the brain.

STRESSED HIPPOPOTAMUS

Not quite a hippopotamus, but close. Some years ago scientists tested animals to see what would happen if they had a damaged hippocampus (a horseshoe-shaped brain area in roughly the middle of the brain). The hippocampus is most famous for being involved in encoding memories.

They allowed the animals to explore their surroundings and then gave them a dose of radiation aimed at their hippocampus. Once placed back into their surroundings they seemed just to base themselves there. Previously they had been enthusiastic about exploring. Now they seemed to have lost their curiosity.

The hippocampus is involved in processing novel situations and things. When the hippocampus was eradicated the animals stopped themselves having these new experiences. This links to how we often feel when we are stressed out – we just want to retreat into a safe secure place.

Hippo attack

The glucocorticoids (a type of steroid hormone), which are released during emotional outbursts and chronic stress, actually break down the neurons in our hippocampus. When Jessie feels stressed she craves familiarity. For other people they may just want routine, which could be their day-to-day normal experiences. Unfortunately, for many people these normal experiences cause them stress. So their way of avoiding stress is actually stressful. It is important to address things that are causing chronic stress.

Inside the hippocampus something quite clever happens. It happens in other areas of the brain and body too, and quite actively in the hippocampus. New neurons are produced, a process called neurogenesis. This means that if we can stimulate the hippocampus, say with novelty (doing something new) then it will plump out a bit and become healthier. Interestingly, a recent study showed that it took around one month for the antidepressant Prozac to elevate a depressed person's mood. This is the same time it takes for neurogenesis to occur.

Jessie starts to realize that how she has been feeling has been natural. Part of her had been feeling that entrepreneurs were all full of life and seeking adventure all the time. She just wasn't feeling like that so was questioning whether she was a good entrepreneur. Now she saw that her craving for familiarity was normal based on her stress levels. She also saw that part of the solution was to actually push herself to do something new to stimulate her hippocampus.

Mirror neurons

As Stuart gets deeper into the coaching session with Jessie he senses that she is open to exploring something that has some pretty 'way out there' connotations for those who want to see them. The research comes from Italy where scientists discovered something called mirror neurons. The story goes that a monkey was wired up for another experiment that was taking place. Then during lunch break a researcher came into the laboratory still eating an ice cream. The monkey's brain responded as if he were eating the ice cream. This was unusual because previously it was thought that neurons should only fire if the thing was happening to the subject. As it turns out monkeys, and humans, can have responses to things going on around them, as if they are happening to them or they're doing them.

When the PA is gently trying to explain to Jessie her mistake, we don't know what is going through the PA's head. Perhaps, underneath what she is portraying, she is actually thinking 'Why can't my boss get the simple things right?' It's conceivable that Jessie is picking up on this, through mirror neurons, and then it sparks all her own thoughts about not being good enough. It may be though, that with the stress she just connects with all her old worries.

In some jobs people feel able openly to connect and share things with their colleagues. They can trust their colleagues to see the best in them. For example, Jessie could have felt able to talk to her PA about how she was feeling pressured, challenged and as if she wasn't good enough. But more often than not people don't feel able to talk to colleagues. Jessie could be seen as weak, as incapable or as struggling. All of which are not good in her work environment. The sad thing about this is that talking, especially for women, helps. Women often use talking to process their thoughts and things often become clearer, less challenging and lose their negative emotional power after talking things through. Men may process better internally so talking is sometimes an unwelcome distraction.

When, during a stress response, adrenaline and cortisol are released and the blood pressure increases in preparation for the fight or flight response, men and women react differently. The male brain at this point becomes agitated and the way it gets its release is through competition, debating or escape. Evolutionarily the male brain was conditioned to be very focused and linear in processing. Goal-oriented direct action helps the male use up the excess adrenaline. The female brain on the other hand remembers the details of what has happened and wants to talk it all through. This way her brain releases oxytocin and serotonin. Serotonin is calming and oxytocin helps her feel close to people, so not alone.

Possibilities

We're moving now to look at how to work with the brain. What can Jessie do to get out of her chronic stress cycle? How can she get rid of that heavy allostatic load?

The first component we're going to deconstruct is how Jessie is finding things acutely stressful. We'll look at what determines her finding something stressful at a particular moment, whereas Judith doesn't seem affected at all by it. One of the biggest contributions here is how Jessie is labelling things. Our narrative and programming attaches meaning to things. This gives us all an opportunity. For Jessie, when the canteen runs out of her favourite breakfast cereal she thinks things like 'Nothing goes my way' or 'The universe is against me'. When she walks from the car park to her office and she steps in a puddle she says 'Why does this always happen to me?'

When she is at work and she makes a mistake she finds herself thinking that people are noticing every mistake she makes and that they know she isn't meant to be here. She was lucky to have started this business; lucky to have made the connections she's made and lucky to have a job. One day they'll figure it out and she'll be cast out. This is a pretty extreme way of thinking about things, but not that uncommon unfortunately. For some people they have hang-ups about their intelligence, for others their looks, others still their personality.

The classic example of someone who overcame the identity people offered him is Richard Branson. Having dyslexia at school he recalls as being a nightmare. He was repeatedly told to do things that really challenged him and embarrassed him. The programming they were trying to give him could have shaped him in a way to limit his future and what he believed he was capable of. His headmaster said 'I predict that you will either go to prison or become a millionaire.' Branson chose to label his experiences in a way that strengthened him. Perhaps hearing a teacher say something like 'You're not clever, or you'll never achieve anything unless you buckle down and learn this stuff' spurred him on to prove them wrong.

So almost everyone is predisposed to label things in a certain way. One of the roles of a good coach is to identify those predisposed labelling tendencies and bring them to your attention so you can consciously change them if they are producing undesirable results. None of your programming is set for life. Through the wonder of neuroplasticity we can change programming to free you up to enable you to achieve your outcomes.

Changing labels is best done over a period of time. It can and does happen over night, naturally – for example when a person is given a job that they never dreamt they would be offered, their view of themselves is turned on its head. Normally though, we need to put effort into changing how we see ourselves.

Jessie could work with three easy steps:

- Decide what new label she wants, for example instead of 'The universe is against me' perhaps 'I'm lucky when it counts'.

- Reflect daily searching for evidence of this new label. Jessie may start noticing that on the day of an important meeting the rain stopped when she needed to walk from her car to the restaurant, meaning her hair still looked good when she arrived. Or she might note that she saved her work just before her computer crashed.

- After two to four weeks she could generalize the label and link it to other labels – she could now generalize to 'I'm lucky' and 'People believe I see the best in situations' and 'People like being around me..

These steps all start rewiring and strengthening new labels that are more useful than previous ones.

BIRD POO AND I

When we label things we need to be honest but we also need to be open to possibilities. For example, recently I was walking through the centre of Birmingham with my mum and husband. We'd finished a little shopping and were heading back to the car to drive out to a nice pub in Harborne. A bird pooped on my head. Just out of nowhere (I realize they don't normally ask first) it pooped on my head. My reaction was to double over in laughter. Eventually a few tears escaped through laughing so hard. My husband was actually in a shop at the time buying some water to drink, so when he came out to see me and my mum in fits of giggles he wasn't sure what had happened. Through the giggles I managed to stammer that a bird had pooped on me, and I'd always wondered what you do in this kind of situation.

I could have been terribly embarrassed, felt really angry with the bird, felt disappointed that my nice clean hair now wouldn't look so good at the pub or even felt like the world was against me. And to be honest I did feel a twinge of 'Oh, I have bird poo in my hair now – that's a bit icky!' But my overwhelming feeling was that this would be a moment to remember, I'd always wondered what one does and now I was going to find out and what an opportunity for releasing some endorphins with side-splitting laughter.

Disclaimer: eventually I did stop laughing and that bottle of water came in very handy to clean me up.

Lack of control

Control, or a perceived lack of it, can be very stressful. Jessie feels that her colleagues will be thinking badly of her and she cannot control their thoughts. She also wishes she could have a holiday, but feels that as the boss she should be here to drive everything forward. This feeling of helplessness has been proven in many different ways to contribute to worry, self-doubt and digressions. These thought processes are energy sapping and make it more difficult to think clearly and complete tasks. At the other end of the spectrum, people who are in control tend to have self-talk that is more focused on problem solving, what they need to do next, what might work, what they are learning.

Stress busting

Key things you can do to reduce stress include the following.

Recognize and label it

Seeing your state and being able to give it a one- or two-word label is very empowering. It can help you get a little distance from the state and experiments tell us that this is a great first step to reducing it.

Reframe

You can reframe things at various stages in various ways. Jessie often feels that her clients have control over her time and can work her like a puppet. Instead, Jessie could choose to see these clients as people she sets boundaries with so they appreciate her more. She wants to serve a client really well and so it is vital that she maintains all the other things that are important to her (such as a holiday, her other clients and her staff) so she can best serve this client. (The key with reframes is to come up with something that works for you. If someone else says it, it can seem a bit far-fetched.)

Socialize

Connecting with other people can help release oxytocin, which can counter some of the negative effects of stress. Although the last thing Jessie feels like doing is going out to see friends, she almost always feels better once she is out and with people.

Contribute

Contributing to something worthy can give you a huge boost. The brain areas that are activated when you do something such as volunteering are the same areas as when you receive a financial reward. You release dopamine and this can help reduce stress. Jessie doesn't feel that she has time to do any volunteering, but comes up with a reframe. She realizes that she often works for an extra couple of hours a couple of times a week. Normally this is following up with a particular client or chasing up some supplier. She isn't paid for this extra time so she is choosing to see it as time she is volunteering. For other people this reframe wouldn't work so they would need to reshuffle their commitments to make time to ensure that they are contributing to something important to them. Some people, for example doctors, get this from their work itself.

Exercise

There is some controversy about the state you do exercise in and whether or not it is then helpful to you, so to reduce stress for sure choose an exercise such as yoga. Overall, the positive effects are great though, so if you enjoy boxercise and it makes you feel better... go for it!

Focus

Stuart checks how Jessie is doing, to ask whether she needs a break or wants to keep going. They decide to stretch their legs and get some water before carrying on. Stress is a huge area and to really get into it can be quite intense. The next area Stuart wants to introduce Jessie to is focus. Our focus is hugely important and shapes a lot of our experience. Sometimes we may feel that our focus is out of our control but this isn't actually the case.

When we are looking at what is possible sometimes it is helpful to go to the extreme to see what is happening there. People with obsessive compulsive disorder are categorized as having an anxiety disorder. They feel completely out of control. Sufferers are plagued by thoughts, intrusive thoughts, that make them feel very uncomfortable and that come with a drive to perform some form of activity to reduce the level of discomfort. At times they focus on their compulsions to the exclusion of everything else. For example, a person may feel consumed by the thought that something terrible will happen to members of their family unless they turn the lights on and off 17 times before leaving a room. Or they may feel incredibly worried about the germs that they believe are all over their hands and need to wash their hands four times an hour. Often the sufferers know that the thoughts are irrational. This can be pretty tough – realizing that your thoughts are crazy – but still feeling compelled to act. We find out more about the structure of this extreme later.

STAPP'S SECRET

An extremely clever physicist called Henry Stapp can help expand our reality here. Stapp believes that 'the conscious intentions of a human being can influence the activities of his brain'. He believes the brain is quantum in its nature. This means that what we consciously think about has an impact on what goes on in our brain. When we think of it like that it sounds quite straightforward. Stapp goes on to say that we control what we think. This is a very unusual school of thought as most people believe the brain adheres only to classical physics. If Stapp is right though, the very nature of the brain as we understand it will be updated. It will be even more important for people to learn to control their thoughts, because no one else is going to. Sir Roger Penrose also believes that quantum processes are important to understand the brain.

Jessie had been on a communication skills course at university because she had wanted to ensure that she didn't end up like some of the controlling and disagreeable bosses out there. As a junior in her first job out of university she'd seen the effect poor communication would have on her colleagues and she'd often be the person comforting them and explaining what the boss actually meant. She remembered that on that course they'd said it was important to take responsibility for your communication and that what you thought had an impact on how your message came across. Today she was realizing that things went a lot deeper.

Stuart knows that there is a lot more to explore with Jessie. To realize fully the impact of the next session he needs to introduce her to some big ideas and experiments slowly. There is controversy over what he is about to share with Jessie but in 20 years we will look back and be amazed that it wasn't being more widely used. We need to understand how brain cells communicate, and a pivotal quantum physics experiment called the double-slit experiment. Without looking at this we are missing out the most fundamental science that, over time, will drastically change how people think.

DOUBLE-SLIT EXPERIMENT

If you read about this experiment and aren't questioning your model of reality then read it again! If it confuses you then you have understood it. Go to the MYBW website at www.synapticpotential.com (archived at https://perma.cc/7875-2HMQ) to watch a clip that helps clarify it further. Richard Feynman is quoted as saying that all of quantum mechanics can be gleaned from carefully thinking through the implications of this experiment.

Basically this pivotal experiment consists of sending a beam of light through a thin plate that has either one or two slits in it. The fundamentals you need to know are simple:

A) Particles can be considered to have mass, we can think of them as tiny balls. Imagine what happens when you throw a tiny ball covered in paint at a wall, you get a tiny circular imprint on the wall.
1) When particles pass through one slit they make a pattern that is in a line; 2) When particles pass through two slits they make a pattern that is in two lines.

B) Waves do not have mass, instead they can be thought of as oscillations or vibrations. Imagine dropping a coin into a water fountain and watching the ripple of waves emanate out from the central point.
3) When waves pass through one slit they make a pattern that is in a line (similar but not identical to the particle); 4) When waves pass

through two slits they make a very different pattern: multiple lines (due to something called interference).

It was believed that light was made up of particles only, discrete balls of matter. The physics-shaking shock occurred when a particle was passed through one slit and it made the pattern in a line (expected) but then went through two slits and it made the multiple line patterns (as in fundamental 4 above) which was very unexpected). This meant that the light was behaving like a wave, which was not at all what Isaac Newton had led us to believe.

In 1961 this experiment was done for the first time with something other than light, electrons. Our understanding of electrons is that they are particles, so they should behave the way fundamentals 1 and 2 dictate. They didn't. In fact even relatively big molecules would behave as if they were waves. The magnitude of weirdness is huge at this point. The physicists tried lots of things to work it out but they just kept discovering weirder things. The particle would act as if sometimes it went through just one slit, or just the other, or neither, or both! This is like a single paint ball being fired towards two gaps and splitting into two to pass through both gaps and then rejoining again afterwards. It didn't make any sense.

Not to be outdone by a particle though, the physicists decided to put a measuring device by the slit to see what was happening. You'll never guess what happened (seriously, it's that amazing): the particle started behaving like a particle again (as fundamental 2).

It is said that the observer collapses the wave function simply by observing, which is physics talk for the particle behaving differently depending on who is watching.

Sir John Eccles won a Nobel Prize for helping us understand how neural communication takes place. In 1986 he proposed that the probability of neurotransmitter release depends on quantum mechanical processes. This means that the brain chemicals that we are becoming familiar with (dopamine, serotonin, adrenaline and around 50 others) are released according to some quantum laws.

Some scientists know that an observer, in this case your own mind, can influence these. In grossly simplified terms this implies that the chemicals that flood and impact your body are affected by quantum processes, which are subject to the effect of the observer. You have an impact on the probability of your mind communicating things.

Focus and obsessive compulsive disorder (OCD)

Let's return to look at focus through the effective treatment of OCD. Often we learn a lot about the body and mind through observing when things go wrong. Recently we've been fortunate to learn a lot through what has helped things go more right. Jeffrey Schwartz works with people with OCD. He helps sufferers literally rewire their brains so that they aren't constantly under the distractive and destructive spell of OCD. One of the tools he teaches people he works with is that of refocusing. For people who have such a strong compulsion to turn a light switch on and off 35 times before they leave a room, this is hard.

Doing something engaging for 15 minutes is the key here. Schwartz has shown that this is the amount of time that it tends to take for the brain to engage in healthy alternative behaviours. Schwartz draws some of his inspiration and knowledge from the Buddhist practices of 'wise attention'. The practice of revaluing something to be 'in accordance with the truth' can be very liberating for a person. The very act of 'paying attention' produces real and powerful changes in the brain. It is thought that the mental force alters brain activity.

The very clever William James said:

'Volitional effort is effort of attention.'

'The function of the effort is... to keep affirming and adopting a thought which, if left to itself, would slip away.'

'Effort of attention is thus the essential phenomenon of will.'

Stuart is keen for Jessie really to grasp how important her focus and attention is. They have an indirect and often direct effect on many things including her state and her actions. The neuroscientist Ian Robertson said that attention 'can sculpt brain activity by turning up or down the rate at which particular sets of synapses fire. And since we know that firing a set of synapses again and again makes [them] grow… stronger, it follows that attention is an important ingredient'. Wilful attention can direct your brain to filter out the suppressive effects of distracting signals. For Jessie this means that by wilfully putting her attention onto her title as the director of her entrepreneurial company it will remind her that she has created this company and she is capable. This attention (which we think of as coming from the mind) then actively changes her brain. Her ability to think that she will get found out will decrease; her brain just won't be wired like that any more.

Action

Stuart senses that Jessie has absorbed as much as she can at this point and that they'll need to return to the quantum components of focus another time. Jessie decides she will focus on her state until the next time they speak. Knowing now the impact it has on her, she is taking the way she interprets and labels things more seriously. Determined to get out of the chronic state of stress she believes herself to be in, she commits to doing some research on meditation and mindfulness. It always sounded like a great way to waste 20 minutes a day when she heard other people talk about it, but if there is some scientific research that substantiates its benefits then she'll consider it. Although she can't really explain why, she also feels as if this is the reality that she has chosen. Things aren't as bad really as they often feel and so she is going to work on ways to keep connected to that feeling.

Jessie's action list

- Commit to addressing the chronic stress situation by doing one new thing a week.
  First week: invest in a yoga class; second week: find a meditation app download and listen to it for five minutes a day; third week: find a dance class; fourth week: call one old friend.

- Go to yoga classes for four weeks. If the investment of time isn't worth it then re-evaluate, possibly get a DVD and do it at home for just 30 minutes a couple of times a week.

- Set an intention to start becoming aware of her negative thoughts for one week to get an overview of the ones that keep showing up.

- The following week set the intention to focus on her positive thoughts and get an overview of her perceived strengths in thought.

- In the third and fourth weeks experiment with reframing negative labels into positive ones (eg 'The heavy traffic is holding me up' becomes 'The heavy traffic is giving me space to think, enjoy the music and plan the rest of my day.'.

- Research setting up a social enterprise arm of the business to give her and her employees even more purpose and sense of contribution in what they do.

MYBW top tips for when something goes wrong

- Be very aware of how you are interpreting things.

- Recognize that failing at something can be hugely valuable – you learn precious distinctions and lessons that may not be gained any other way.

- Feeling guilty can activate the same areas of the brain as physical pain – so use guilt carefully.

- Reframe, revalue and do everything you can to position yourself as in control of your life.

- Practise exerting your focus in different ways to strengthen different neural pathways.

- Decide how you want your brain to respond when something goes wrong.

- Increase your ability to place and hold your attention where it best serves you.

MYBW top benefits for mastering things when something goes wrong

- Free yourself up from experiencing negative knock-on consequences.
- See new opportunities where previously you saw only problems.
- Control your world from the inside out.

3
Learning how to influence what feels out of your control

Ben picked up the phone for his third call this month with his coach Stuart. It had been a tough day and he still had more to get done before he could go home. Not that he was especially looking forward to going home either. Things with his wife Rebecca weren't exactly going well at the moment.

As he heard Stuart's voice on the other end of the phone he felt his shoulders relax slightly. Although he wasn't sure what he would work on with his coach today, he knew that for the three months he'd been working with Stuart, they always made progress and he felt better after the calls. Ben started to recount the day's events, starting with arriving at work to see that the trainee accountant, Jane, who was doing her placement with him currently, had failed once again to follow his instructions. When Jane got into the office he had practically yelled at her, saying that if she didn't learn to follow simple instructions she'd never make it at this firm. Although he did apologize later, she was still being slow with everything she had to do.

As Ben started recounting other things that had frustrated him that day Stuart interrupted him and asked for them to check some general trends in Ben's life at the moment. For the first time, they talked in depth about his relationship with Rebecca. It wasn't that there was anything hugely wrong; they just didn't seem as close as they used to be. She seemed to be snapping at him more and was often too tired for sex.

Stuart asked Ben to take a step back from his life and look for the general things that were occurring. Thankfully Ben is quite self-aware – with other people much more questioning would probably have been required. Ben observed that he gets irritated at things easily. One of his colleagues, Mark, never washes his own cup up in the kitchen and every time he sees this guy he just thinks what an arrogant man he is. In meetings he finds himself frowning every time the guy speaks. Ben also realized he is getting frustrated, almost angry, on a daily basis. It seems to him as if he is working with a bunch of people who cannot seem to get it right. No one seems to have passion any more, including himself.

Situation

Stuart explains to Ben that he needs to understand what is going on inside that brain and mind and even body of his so that he can have more control over how each day goes for him. The extended benefits of this are that his relationship with his wife should improve and the people he works with will become more productive. Overall, the quality of each day will substantially increase. There is a lot to cover to get to that point so today they will just focus on him and then next time look at how this works to affect other people in more depth.

This chapter is about handling your negative states to enjoy life and your relationships more. The bonus is getting better responses from other people.

States, emotions and feelings

These terms are normally used interchangeably but that is normally by people who understand very little about them. The distinctions between them are, in places, fine, but in order to gain better control over them a good place to start is to gain an overview of what they are. Emotions are made up of internal experiences and involve neural and chemical processes. States are highly generalized ways of being. Feelings are your perception of the changes occurring internally. The words we use to describe our specific states, emotions and feelings are the same. For example, we can be in a state of excitement, have the emotions of excitement or feel excited. You could possibly experience all three at once, but also possibly not. You could have the emotions of excitement but feel anxious. You could be in a state of heightened awareness and feel alert and have the emotions of calm and alertness.

Both thinking and emotions are hugely valuable to us as humans. Thinking allows us to process things and play with different ideas. Emotions are designed to precede action. Emotions cause physical changes in your body and prepare you to do something. Emotions almost take over your brain; they are absorbing and have a series of powerful effects.

> All emotions have the potential to cascade to generate feelings when you are awake and alert, not all feelings originate in emotions.

To know or not to know

Imagine a scenario where Ben found himself working on a team with Mark, the guy who never washes his cup up. At some point Ben becomes aware that he is disagreeing with almost all of Mark's points while a group of other people seem to be agreeing with the guy. The spiral continues until Ben is feeling thoroughly dejected because it's as if he is the only one who can see through this guy.

A shift in thinking and feeling is unlikely to happen at this moment, when Ben is feeling low, but it could happen later. With favourable circumstances, such as being relaxed and feeling safe, he could start to explore what is really going on with Mark, which is much deeper. (We discover exactly what later in this chapter.) This would give him the edge and access to valuable information he would need to use to be as successful as he can be.

There are links here to the psychological concepts of cognitive bias. We have been exploring one of the potential neuroscientific underpinnings to this phenomenon.

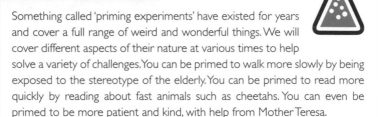

HOW TO WIN A GAME OF TRIVIAL PURSUIT

Something called 'priming experiments' have existed for years and cover a full range of weird and wonderful things. We will cover different aspects of their nature at various times to help solve a variety of challenges. You can be primed to walk more slowly by being exposed to the stereotype of the elderly. You can be primed to read more quickly by reading about fast animals such as cheetahs. You can even be primed to be more patient and kind, with help from Mother Teresa.

Two scientists did an experiment to investigate a link between stereotypes priming on overt behaviour. This involved priming people with the stereotype of a professor, a secretary and no priming at all. In this case the priming was done by getting people to read something and there was either strong mentions of a professor or a secretary or neither. The idea was that professors are thought of as knowledgeable and intelligent whereas secretaries are not. The people were then asked to answer multiple-choice questions – from the game trivial pursuit.

The professor-primed people answered more questions correctly than the other two groups and the secretary-primed people answered the questions most quickly.

The implications of this experiment are huge. Using a very straightforward method (reading about something) we have managed to make people more intelligent or quicker than they would otherwise be. We read things every day, how are they priming us to behave? Reading is only one way of priming people. Listening to things, such as music, is another powerful one. You can probably imagine or remember the difference it makes walking into a party after listening to upbeat party music versus listening to some sombre classical music. Your body language, thoughts and energy are all very different. The list of what can prime you is limited only by your imagination.

So what can you do to really use the opportunity priming gives you? Questions anyone who wants to take more control needs to ask themselves are:

- What do I want to achieve?
- What do I want to improve in my life?
- Who excels at a component of that?
- How can I most easily prime myself?

For example, Ben worked through these questions and realized that it was important to him to have a healthy relationship with his wife. In order to do that, he thought he needed to be more patient with her so he decided he wanted to improve his patience. When he looked at the third question a person jumped into his mind but he wasn't sure it would work with him. Stuart said he could give it a go. Ben had thought of Jesus because he remembered being taught that he was very patient, kind and gentle. When he thought of him he got a picture of him moving through crowds of people and people wanting his attention but him still remaining calm. He felt a bit like this at the end of a hard day's work when his wife just wanted to talk about everything. Considering how he could most easily prime himself he decided upon listening to a calm song on the way home, and reading the proverb 'A hot-tempered man stirs up dissention, but a patient man calms a quarrel.' He'd also picture Jesus being calm despite many people wanting his attention.

Controlling our feelings

Can we control our feelings? The short answer is 'Yes', the long answer is 'it's a challenge'. There are several causes of feelings. The most well known is the variety of feelings that occur when we are in a particular emotional state. When our body is flooded with a certain selection of chemicals we experience the associated feeling. For example, when we have the happiness 'prescription' of emotions then we get that happy feeling.

Another cause of feelings is background feelings. We experience these a large proportion of the time. As their name suggests, they exist without drawing too much attention to themselves. They don't tend to be memorable, they're just there. These feelings can be changed by changing your emotional state and thus the feelings linked to those.

Emotional state = apprehension → Feelings = nervous, unsure, negative

Emotional state = excited → Feelings = anticipation, motivated, positive

Feelings that exist for a long time or that are fairly frequent can be considered moods. These background feelings are distinctly different from the more powerful feelings associated with emotional states.

It is useful to be able to distinguish between feelings because a higher level of self-awareness has been shown to lead to better leadership of self and others.

Controlling our emotions

Emotions certainly are the big players in terms of the effects they can have on our lives. Ben is currently finding challenges all over his life as a result of his emotional states. So what is happening that is so powerful?

JUST STOP IT

It used to be commonly espoused that our emotional responses were out of our control. That some people were just angry people, or frustrated people. That our amygdala was subject to hijacking at any time. This kind of thinking is how it can appear from the outside. In the moment it certainly can feel like our amygdala is in control and our prefrontal cortex is powerless.

However, research has moved on. It is now more understood that emotions are not an objective thing. Instead they are learned and our brain constructs them. Our brain is always trying to regulate the body. It predicts what is needed. It relies on data and experience to allocate attention and other precious resources.

It is not the case that we can 'just stop it' when it comes to experiencing any emotion. In the moment they feel very real and indeed are. However, over time the emotional responses can be changed.

Practically possible

Ben asks if all this means that he should just not get angry with Jane, that he should be sympathetic rather than angry towards her. Stuart says that yes, in an ideal world that would be the response that over time would get better results from Jane and from him. However, 'just' doing anything in regards to emotions or feelings isn't easy. The reason it isn't easy though is just because most people don't understand it. It's like saying to a wedding cake maker 'Just make me a cake' – that's fine for experienced people who know what ingredients go into the cake, what tins are best to bake it in, how to line the tins so they can get the cake out afterwards. They also know how their cooker works, how long they need to bake it for, how to test when it's done, how to cool it effectively and when it can then be assembled. They know what structural equipment they need to assemble the cake and then how to decorate it. (We could go on but you get the picture that there is a lot to know about how to make a wedding cake successfully.)

If an inexperienced person planned to make a cake, they may think they'd 'just' make it. They'd soon realize that there are a lot of components that you need to get right in order to make a cake successfully. The same thing is true of managing emotional states. Some things you need to take account of are:

- priming;
- anchoring;
- focus;

- beliefs;
- values;
- intention;
- reprogramming.

New for Ben

Stuart starts to explain to Ben that our concepts of emotions are learnt. The experiences are typically narrated to us by parents or other inputs and this shapes our understanding of them. When we hear a term for an emotion and link that to an experience it becomes easier to construct that again and again. Different cultures have different words for different concepts. For example in Tahiti they don't have 'sadness'. They do have a word for 'the kind of fatigue you feel when you have the flu' which isn't the same as sadness, but is what they would feel in similar situations to us.

Increasing our emotional vocabulary can help us construct a more helpful present. Like many things neurally, once we practice it, the ability becomes stronger. Discerning between different emotions is a skill. By Ben narrating his experiences differently, with the help of his prefrontal cortex, he can literally reprogramme his emotional responses.

Stuart decided to move Ben onto a simpler concept with practical things he can do quickly. Anchoring is one of the biggest contributions to poor productivity that exists, and is certainly a problem for Ben both in his personal and professional life. To understand anchoring, we need to understand how the brain communicates within itself.

THE NEURONAL STORY

The brain is filled with cells called neurons. The ends of the neurons don't actually touch, but form a gap, called the synaptic gap. They communicate to one another through electrical or chemical signals that pass across this gap between them. The first time a series of neurons communicate a message with one another it's as if a very faint pathway is made. The next time that same message needs to get through it is slightly easier because there is that faint pathway. The more times the message is conveyed the easier it gets for the message to get through. We can think of this as a deep groove being created. When a new message that is linked to the old message starts to be communicated it may end up linking into previously wired neurons. For example, if every day you see a cup being held by a person you dislike, your 'map' for that person and that cup will be well interlinked. You will have an emotional response that will be triggered by neurons communicating to one another causing the release of hormones from various places. If you were to then see the cup without the person, the same 'map' or 'engram' would be triggered and you'd end up feeling the same.

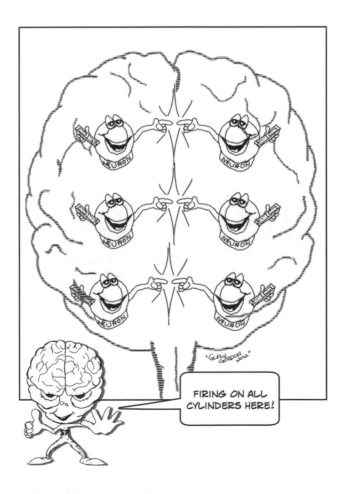

Henry David Thoreau says: 'As a single footstep will not make a path on the earth, so a single thought will not make a pathway in the mind. To make a deep physical path, we walk again and again. To make a deep mental path, we must think over and over the kind of thoughts we wish to dominate our lives.'

Why anchoring is vital

We can now see how one thought can spark many other thoughts. One reason anchoring is so powerful is that it happens naturally and unconsciously. One neural net gets linked into another and another and before you know it you are feeling something you weren't expecting to. This happens many times a day. Often it can be useful. For example, some people have anchored themselves to the emotional state of 'raring to go' when they step into the office. They feel full of energy and ready to deal with whatever the work day could throw at them. When they pour a cup of coffee they start to experience the effects of the caffeine perking them up (before they've even taken a sip).

Anchoring can be harnessed and consciously applied. For example, those same people could at lunchtime always step outside for five minutes and breath deeply the city air. They've conditioned themselves to trigger a flood of calming chemicals enabling them to enjoy a much-needed change of emotional pace. As soon as they step back into the office they feel revived and ready for the next challenge. At first they have to work at this routine by thinking calming thoughts when they step outside, but pretty quickly (certainly within a few weeks) it will happen automatically without them even having to think about it.

About the cup

Ben thinks his issue with Mark is about his cup, well, what the cup represents. Interestingly, his great friend Elaine never washes her cup up either, but he still really likes her. The cup is actually just what Ben has consciously told himself the problem stems from to make his feelings logical.

So what happened with Mark that has tainted every subsequent interaction? The day that Mark joined the team Ben had received a very traumatic phone call. He had been told that his father had just passed away in the nursing home and that there would be an investigation into how it had happened. It was thought that one of the staff members had been under the influence of alcohol at work and had made an error with the medication. Ben was understandably in a heightened state of anger feeling that incompetence was to blame. He walked out of his office and almost walked into Mark and his boss. He was introduced to Mark and as he took a deep breath to calm himself he inhaled the strong aftershave Mark was wearing.

Neurologically the intense state Ben was in became linked to the smell of the aftershave and the visual impression of Mark. This became wired in to Ben's neurology. The next time Ben saw Mark, in the kitchen, he smelt that same aftershave and felt those feelings of anger again. He felt them more mildly of course this time, but they were there. He also felt that the guy was incompetent. Consciously this didn't make any sense, this was only the second time of seeing Mark, and he couldn't say he knew him at all. So his brain came up with a reason that was plausible to him as to why he was feeling these things. The cup was the only real thing available.

By the time Ben saw Mark in a meeting he had his feelings about him pretty well established, he'd only seen him two further times and both times he had a cup in his hand. In Ben's book Mark was now arrogant, thought he was better than everyone else, didn't really know what he was doing and needed to be kept in check. This was a mix of what Ben had been programmed to think incompetence meant (as a result of the initial cross over of emotion he felt about his father's carer to Mark) and what he generalized not washing your cup up meant.

In their first meeting Ben, unconsciously, needed to substantiate his beliefs about who Mark was. This meant that he was primed to seek out the evidence that supported this. He wouldn't hear any good ideas Mark put forward – instead interpreting this as arrogance. He would hear many bad ideas – rather than seeing that Mark just didn't know they'd tried that before, since he was new to the firm.

By understanding how anchoring works you become better equipped to spot when it is at play. You are also able to open your mind up to the possibility that what you initially think about someone may not be the whole story.

Here are a series of experiments that substantiate the depth of anchoring. Remembering, raising your self-awareness of how you are being anchored and how you are anchoring others increases your ability to lead yourself and others.

THE IMPORTANCE OF ATTENTION

In the 1970s Drs Corteen and Wood ran some experiments that involved electric shocks. The participants were given these electric shocks when they heard the name of a certain city. Their bodies responded with a fear response, which was measured by a sweat sensitive pad on their skin.

Next they were told to listen to words spoken into one ear and to repeat out loud these words. There were also other words being spoken into their other ear but they were instructed to ignore these. The result was that a fear response was still detected when the city name was said into that ear even though they weren't consciously listening to what was being said.

SEEING

A fantastic set of experiments showed that seeing isn't even necessary for triggering responses. The experiment consisted of showing people a series of pictures of human faces. They were shown very quickly and occasionally a loud burst of unpleasant noise sounded. The people's bodies responded with mild fear, unsurprisingly. After a while the people had become conditioned to respond with fear when they saw certain faces. This was all to be expected and has been proven in many other experiments. Where it got extra interesting is when the researchers showed the pictures so rapidly that the people couldn't consciously perceive them. The result was that some areas of the brain didn't in fact respond but the amygdala had the same response. It has been described as being as if the conditioning portion of the experiment sets the amygdala's sensitivity at a higher level.

The brain is constantly recording things. Positive experiences are recorded through chemical signatures of neural nets as an episodic memory. (Episodic memories are autobiographical – the things that have happened.) When we trigger that memory in any way, for example by seeing a face, the emotional state and the feelings associated with the neural net is experienced again. Technically your body responds as if the experience is happening again. That is why stress reduction techniques recommend remembering a peaceful place – because your body will flood your body with those peaceful chemicals again.

REMEMBERING

Things get a little weird here because the following story illustrates that we don't even need to be able to remember consciously what has happened to have an emotional response to it.

In the early 1900s a French physician called Edousard Claparede was looking after a patient who had lost the ability to create new memories. She had brain damage that meant even when Claparede left the room for a few minutes when he returned she wouldn't remember him.

Every time he went into the room he would shake her hand and introduce himself to her. One day he decided to do something different. He concealed a tack in his hand when he shook hers. She had a typical fear response and pulled her hand away. This was to be expected. However when Claparede came back next time he offered his hand to her as normal but she refused to shake it. She couldn't tell them why she wouldn't shake his hand but she wouldn't!

The brain stores information – it doesn't know what is useful or not, it just stores it and tries to protect us.

There may be times when you have an instinctive reaction to something, but you can't explain why. It is a brave (but perhaps uninformed) person who blindly ignores their reaction and presumes they are being illogical. Had Claparede done this she could have been pricked with a tack every day for the rest of her life.

The importance of smell

Smell is very important because of how quickly it gets to the brain. All other sensory information takes time to get into the brain (thalamus, cortex then onto limbic system) whereas smell goes direct to the limbic system then onto the cortex. Studies have shown that smell enhances memories. One study got students to learn new words while sniffing an unusual smell. They were then able to sniff the unusual smell again when trying to recall the words. The results were 20 per cent higher for the group that used smells. It is also interesting to note that women have a much more sensitive sense of smell than men.

Action

Now Ben understands that when he turns up to work after a tense start to the day with Rebecca he needs to change his state. If he doesn't then he is primed to be irritable and easily frustrated and mean to people, for example Jane, at work. Ben decided to play one of his favourite songs as soon as he got into the car to drive to work in the mornings and not to put on his cologne until he got to work. During the drive he shifts his attention from home to work and thinks about what he enjoys about his job and the sense of satisfaction he gets from completing projects.

Ben's action list

Ben knows that he works best when he focuses on just a few simple things so he chooses to:

- Listen to music in the car on the way to work.

- Focus on things he enjoys about work while driving in.
- Put cologne on in the car park at work (to anchor in the start of a focused new work day).

He also puts these things into his phone as a list and looks at it every morning. That way he knows he'll stay on top of them and keep them in his awareness.

MYBW top tips for negative states

By understanding how your brain actually works you are able to come up with the best ways for you to:

- Authentically seek to understand other people's emotional responses to you.
- Prime yourself to experience positive background feelings.
- Remember: 'cells that fire together, wire together' so be aware of your regular thoughts.
- Positively anchor smells that make it easy to get into positive states (such as a favourite perfume or a great wine or flower).
- Critically analyse from a different perspective your opinions of others. Do these opinions hold up to scrutiny? Would your life be easier if you gave people a clean sheet and focused on their attributes.
- Become aware of negative anchors and change them.

MYBW top benefits for mastering negative states:

- Form more fulfilling relationships.
- Enjoy life more.
- Get better responses from other people.
- Feel more in control of what happens.

4

The challenge of being everything to everyone

The hidden depth and revealed simplicity of time management

Kate wasn't sure she was ready for this week. It was already Wednesday and she felt behind. Six months ago she had taken a time management course and, as she explained to Stuart, she knows the theory. The idea of putting first things first was magical to her when she first heard it. It made complete sense and she really felt it was something she could do.

Kate has her personal mission statement – she knows what her values are, her roles are and what the most important things are to her this year. She had found, after last month, that she preferred to write her plan the evening before so she could be thinking about things before she got to work in the morning. This week she had wanted to write a full weekly plan on Sunday evening but was exhausted and then her friend called so she spoke to her for about an hour. By the time she'd finished it was late and she knew she needed to sleep.

On Monday morning Kate had intended to write out her full plan for the week. She would make sure the most important things (for all her different roles) had space in her diary. There would be room left for all the other things that would no doubt crop up. It didn't quite happen like that though. Kate wrote down to do some preparation for her promotion interview for Monday evening and two important work tasks for during the day on Monday and to have lunch with her fiancé on Tuesday. That was as far as she got for the week though. Not exactly a full plan.

When questioned Kate revealed that she didn't get to do her preparation on Monday evening because a friend texted her during the day and asked if she would go round and help her paint her soon-to-be baby's room. People are what Kate finds it hardest to say no to. If someone asks her to do something, or even asks her if she is free she'll say 'yes' unless she physically can't be there because she is seeing someone else.

Nor did Kate do the second important work task because a big pile of papers arrived for a case she was seeing the client for on Tuesday so she had to look at them. On Tuesday she did meet her fiancé for lunch – he gets really cross if she cancels things they've planned together.

So now it's Wednesday and Kate is feeling pressurized and she has abandoned her plan since 'it wasn't working this week anyway'. Also, since her planning didn't extend past Tuesday lunchtime it was out of date now anyway. As fast as she decides to work on one task the phone rings, another e-mail lands in her inbox, because she has forgotten to switch them both off, someone pops their head around her cubical to ask her a question, a friend texts her and she remembers something she hasn't done yet. As if that isn't enough in the back of her mind she keeps getting flashes of things that need organizing for her upcoming wedding.

Finally, she confides in Stuart something that they don't normally talk about, but she feels she can. She is struggling to stick to her exercise plan. She knows it is important and really wants to feel confident about her body, which she does most when she is hitting the gym three times a week. Fitting that in along with seeing friends, her fiancé, working hard and having some time off just seems impossible.

Situation

It seems Kate is experiencing a range of challenges. So what are the situations that are actually going on for Kate on this Wednesday afternoon? Stuart wanted them to both be aware of some important time management factors:

- She left her weekly planning until Sunday evening.
- She let urgent work take the place of planned work without rescheduling it.
- She abandoned her plan.
- She isn't taking responsibility for working her plan (there isn't a magical genie that jumps out of the plan).
- Friends are extremely important to Kate, and the manifestation of this is that she will reprioritize seeing them.
- Distractions are slowing down her work.
- She is not getting to the gym regularly.
- She is not sure everything is possible.

This chapter is about regaining that lost time to increase your productivity and sense of control. We look at strategically how to minimize distractions, achieve your goals more easily and feel great about yourself. The bonus is that you increase your integrity.

STUDYING DISTRACTIONS

A study by Basex, a New York research firm, found that employees spend an average of 11 minutes on a project before being distracted. It then takes 25 minutes to return to original task, if at all. People switch activities every 3 minutes, either by making a call, speaking with someone live or working on another document.

Another study showed office distractions to eat up an average of 2.1 hours a day. This data highlights a major problem for people's productivity. Losing this amount of time each day really adds up.

Possibilities

The big picture here that Stuart wants Kate to see is that she has half a strategy. She knows the theory of 'what' to do. She has taken the time management course, before that she has read the time management books. The problem is not knowing what to do. When we aren't doing something the way we want to be, it is a natural response to say that it doesn't work. We want to protect ourselves and when we aren't doing something, such as making our weekly plan, then it's easiest to say that the tool isn't working.

Understanding what is going on behind the scenes, actually inside her brain, will give Kate a whole new level of appreciation for what she is trying to do by seeing 'how' her brain is actually doing it. There are three main things to look at from where Kate is at today. Decisions underpin five out of the eight situations she finds herself in. Stuart would normally look at this first with Kate while her attention is at its freshest but he needs to calm her down first. So he chooses to address distractions first, which only accounts for one of the situations she talked about, but the actual time this eats up is huge. The decrease in productivity due to distractions amounts to around 2 hours a day, 5 days a week (that is 10 hours a week, 40 hours a month). Since Kate works roughly 10-hour days that is 20 per cent of her time lost to distractions.

Distractions

One of the biggest problems with distractions is that our brains are wired in a way that makes it very easy for us to get distracted. Structurally one little trigger can set our brain off in an entirely different direction to the one we wanted it to be going in. It's like a big plate of spaghetti where all the squiggles are connected to other squiggles. Everywhere one touches another there is a potential for a distraction to occur.

ENVIRONMENTAL PRUNING

More and more, Synaptic Potential are being called into organizations to help them make their environments functional. Gone are the days where everything just looked beautiful — now we politely demand productive as well as attractive. Here are some top considerations for focused, individual, analytical or highly cognitive type work:

1. What is within a person's visual field is really important.

2. Green plants are helpful in lots of ways

3. Reduce peripheral distractions

4. Consider auditory protection

5. *Remember:* different desired results require different environments

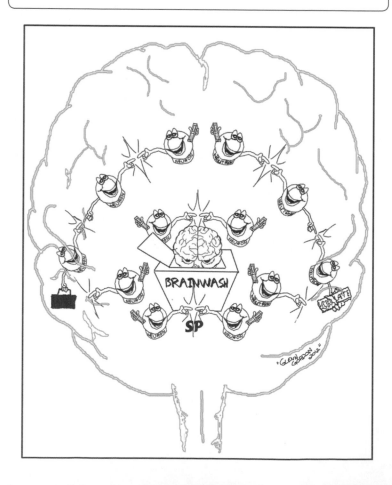

**AMBIENT NEURAL
ACTIVITY = BUSY BRAIN**

Your brain is always on. Data is pouring into your brain through your many senses (your eyes, ears, skin, nose and taste buds) and your brain is constantly processing, reconfiguring, reconnecting trillions of connections each and every moment. This is called 'ambient neural activity'. This is a good thing to be aware of because if you want to prevent distractions then you have to focus your attention. If you don't the ambient activity may come to the forefront of your awareness and distract you.

Distractions deplete your limited prefrontal cortex resources. You'll remember you need your prefrontal cortex for doing anything where you'd say to yourself 'I need to be switched on right now' — such as attending important meetings, checking reports, learning new information or writing documents. Being available constantly by phone, e-mail and text can drop your IQ by as much as losing a night's sleep.

Busy brain in practice

What does this busy brain mean in practice? Well, if Kate had planned what she would do on Monday, and had allowed enough time to do everything by 6.00 pm when she wanted to go to see her friend, it's likely she wouldn't have left work until 8.00 pm. Does that scenario seem familiar? On Monday Kate didn't stay late, but instead didn't get around 20 per cent of the things done that she wanted to get done. On Tuesday the same thing happened, so by mid-week she has around six hours of outstanding work that she wanted to have done but hasn't.

What are her choices? Either she just doesn't do it, or she has to fit it in. Hardly surprising that she is feeling time management doesn't work when she is looking at working until midnight Wednesday evening just to get up to date.

Benefits of braking

Thankfully your brain does have mechanisms to ensure that it doesn't have to follow every white rabbit down a neural rabbit hole. The prefrontal cortex is responsible for 'braking' your thoughts. So your ability to focus isn't just about looking elsewhere, it is also about stopping other thoughts from entering your consciousness. This process of braking thoughts uses a lot of energy and your ability to brake gets depleted each time it is used. Braking really needs to be done before the momentum of a thought takes over. It is a natural process that you are already familiar with. The last time you were working on an important piece of work and the thought 'What shall I have for lunch?' jumped into your mind you had the opportunity to 'brake'. If you said to yourself something

like 'I'll decide on that later, I need to stay focused now' and the thought left your mind then you 'parked' that thought. The benefits of braking are that you are able to be more productive by continuing to work on one thing at a time rather than switching between many.

MYBW top distraction management tips

- Clear your mind first – write things down to empty it, decide on action to be taken to start to solve any problems.
- Remove external distractions when you need to focus (ceremoniously if possible as this can have anchoring benefits too).
- Prioritizing is very energy intensive so do it at a time when you are most switched on.
- Keep a 'tidy or integrity-filled life' – use a coach to keep things in check so they don't distract you (for example a poor relationship or health worries).
- Practise mental 'braking' – calling unhelpful thoughts to a halt and 'parking' thoughts until later.

Getting stuck

Kate has become unsure that doing everything she wants to do is possible. This is a big problem. She is also unable to reconcile her friends needing her, and her feeling that she has to be there for them, with also getting her work done. This is also a big problem. Both problems could have similar root causes in the brain, which we will look at in a moment.

The first thing to validate with Kate though is that valuing her friends is a good thing. It being important to her to be there for her friends isn't the problem. That is all Stuart tells her at this point, he believes these problems have strategic solutions rather than pure brain-based solutions. Strategically Kate could organize her schedule in a way that leaves gaps. The gaps could enable her to move planned things, such as work, around if something urgent or unexpected crops up with her friends. This simple strategic move would have solved the Monday problem of her having promotion preparation work scheduled in the evening but then feeling she should go and help her friend paint. She could have simply swapped the Monday night preparation for Wednesday lunch or evening if she had a gap there.

The other strategic consideration is that she could say no to her friend in a way that still shows she cares and wants to be there for her friend. She wants to prepare for the promotion for a reason, it is important to her too. It brings its own potential benefits and not doing it brings potential negatives. Sometimes the best thing to do is to say 'Thank you so much for inviting me over to help paint, I'd really love to help you out and spend quality time with you. You know how I have that promotion interview coming up, well I've arranged to do some preparatory work for it tonight so that I can feel on top of it this week (I don't have another free evening until Friday) so how would you feel if I came around on Saturday instead? If there is still painting to be done then I can help you or if not I can help you buy or build furniture for the room then? Would that be okay?'

Strategic planning

Whenever Kate has planned a full week out it looks like the schedule of a prime minister. Every hour has something in it and there are no gaps. She almost sees gaps as a missed opportunity, just waiting to be filled by the first friend who asks when they can hang out. Stuart is aiming to transform how Kate looks at her week.

Kate must ensure that she does put first things first in order to be best able to:

- be there for people she cares about;
- be optimally productive;
- serve those who need her;
- do a great job;
- feel good about her week.

With Kate's lifestyle she needs to maintain a degree of flexibility. This may be to respond to visiting a sick friend or taking an important piece of work home to finish. An already full schedule doesn't permit this. Kate needs to realize that being brilliant doesn't mean being all things to all people every single day. Sometimes it is about connecting with people once a month or for just a few, every week.

Any more than this is totally unrealistic for most people and is what lands them in trouble. Stuart is keen not to be too prescriptive; Kate knows the theory, so she needs mobilizing to action it (which is what the rest of the session is about) and also to start with just one strategic change. Leaving a set number of slots during her week free until the start of that week, and only putting in vital things would be that change. If something comes up that enables her to fulfil one of her roles or responsibilities that isn't being fulfilled already that week then she could put it in at the last minute. If a once in a lifetime opportunity (or near enough) comes up, say a last-minute city break with the girls, then she can go knowing that she has gaps in which she can make up anything she was going to do during that weekend. This strategic change is about having fewer things being scheduled in stone, as this leads to more being done overall.

Getting unstuck

Whether a problem has a strategic answer to it or another type of answer (like a values shift or a change in belief) the challenge is often getting to a place where you can find that answer. One of the biggest mental time wasters there is we call the impasse. The impasse is a term borrowed from common use meaning any situation where no progress can be made. In brain terms we use it to describe the situation where you cannot seem to come up with a solution, you are stuck on the same old thoughts.

In Kate's situation she had her deep-rooted need to value her friends. This kept her circling around the thought 'If I'm going to be a good friend I must go and help her tonight'.

Other people's insight's can be hugely valuable to us – in the case of Dodge his ability to think outside the box has since saved lots of

BETWEEN LIFE AND DEATH

Sometimes the inability to get out of repetitive thought patterns is deadly. The best example of this is told beautifully by the science writer Jonah Lehrer in *The Decisive Moment*. It is the story of Wag Dodge's impasse that saved his life and would become a recognized method of fire fighting. Dodge was in charge of the group of fire-fighters that were dispatched to the area of Mann Gulch during one of many major fire alerts. This is a strange area of around three miles where the Rocky Mountains meet the Great Plains, where pine trees meet prairie grass and where there are steep cliffs.

Dodge arrived with his team and immediately took them towards the Missouri River because he didn't trust the blaze and wanted to be near water. He was very aware that the fire could crown, which would mean the flames reached the tree tops. The fire would then spread quickly, with embers in the air, across the prairie. Dodge was also very aware that it was the hottest day ever recorded in that area and they didn't have a map of the terrain. They also didn't have a radio because its parachute had failed. It was 5.00 pm, a dangerous time because the twilight wind can change the direction of the fire quickly. The fire had been blowing away from the river, with the men safely on the other side, but suddenly changed direction. Very quickly the fire was on their side of the river, burning the grass and accelerating towards the men at speeds of 30 miles per hour. The fire burns everything it comes into contact with. The men started running towards the steep cliffs, knowing that once the fire hit the slope it would accelerate. When they started running they had a 200-yard head start. Now they only had 50 yards on the fire, and Dodge realized within 30 seconds it would be upon them. So he stopped and stood still and yelled at his men to do the same. He fought his most basic instinct to escape danger. His men didn't listen or didn't hear him and kept running. In that moment an idea came to him.

He lit a match and burnt a small area of ground. He laid down on this burned land, wet his handkerchief and put it to his mouth. He then waited. After a few minutes he got up, safe and unharmed. 13 of his men died that day. Dodge's escape technique is now standard fire-fighting practice. Most of the men that day kept the same thought circling in their brains; I must run. They had the added survival instinct interfering with their ability to engage their prefrontal cortex to let them think clearly.

lives. Sometimes we need to have our own insights. Kate starts to realize this as it dawns on her that there are other ways she can be a great friend without having to say yes to everything that her friends ask her to do.

To get past an impasse we must let the brain chill out. We have to stop the wrong answers (or same old answers) from circulating repeatedly. Different people find different ways most helpful to do this. Some distract themselves and then let the new answer bubble up. Some actively ask themselves what the other options are. Some consider the opposite answer and work strategically back considering other options.

Having an insight involves hearing subtle signals, allowing loose connections in your brain to be made and become strong enough for you to recognize them. If there are a lot of other things going on then you'll never hear the subtle signals. The more relaxed and happy you are the more frequently insights occur. This is a great business reason to do things you enjoy, such as golf, swimming, eating good food and going to a spa. Investing time and money in things that enable you to come up with genius ideas that you wouldn't otherwise have will quickly prove to be a great investment. Just telling Kate 'You don't have to jump to every request your friends make in order to be a good friend' probably wouldn't have a deep enough impact on her. In order for her to get past that impasse she needs to make the new connections herself. This is where a coach helps her, but cannot do it for her. Here are some tips to help you get yourself unstuck more frequently.

MYBW top 'getting unstuck' tips

- Keep a set number of slots free in your diary to enable you to move things around at the last minute if necessary (experiment to find the right number of slots for you).

- Be realistic about how often you can spend quality time with each of your friends. Are they yearly friends? Monthly friends? Weekly friends? Can you connect with them in different ways between spending quality time with them? Guilt often arises from unrealistic expectations.

- Do enjoyable things. For example, organize a last-minute game of golf or trip to the driving range.

- Call some friends and go to a pub to play a card game while catching up. Choosing some of your favourite activities (that aren't heavy going) gives your brain a break.

- Reduce your pressure. Sometimes people find pressure useful. Experiment with reducing it and see if your brain responds positively.

- Take regular breaks so you are working with a fresh brain.

- Incorporate some quiet time during each day to see if there is anything in the background of your mind that you need to pay attention to.

- When faced with a problem break it up to its biggest components. Look for connections and patterns without trying to solve it.

- Increase your awareness and openness.

Decision making

Kate's initial response when Stuart starts introducing decision making is 'What has that got to do with time management?' This is a great question and as soon as they get into it Kate sees that decision making is fundamental to managing her time.

An overworked frontal lobe exhibits similar challenges. If your frontal lobe is tired, from doing a lot of thinking or complex cognitive work, then you will struggle to make decisions that involve your free will. For example, if you go food shopping after a long day at work you may find yourself staring aimlessly at the choices for dinner, not engaged in making a speedy decision at all.

Brain processes

Kate was interested in what went on in her brain when she made a decision. What was it that happened to make her leave her planning

ENTER DR GOLDBERG

You will meet Elkhonon Goldberg again as he is one of the great thinkers of our time. Currently a professor at the New York School of Medicine, he has advanced our understanding of the frontal lobe and its role in free willed decision-making.

In this set of experiments Goldberg took two groups of people and explained that they would be shown a geometric symbol and then asked to choose one from two additional pictorial designs. They were assured no answer was right or wrong it was simply personal choice. One group were healthy individuals, the other group had various types of brain damage.

As you can guess, the healthy group had no problems choosing a design. Those who had frontal lobe damage in the other group though found it very difficult to decide. (Those who had other types of brain damage could still choose freely.) When Goldberg adjusted the test and asked the people to choose a design 'most similar to the target' (the original geometric shape shown to them) and 'most different from the target' there was no difference in the performance. Those with frontal lobe damage were able to make a decision based on decision, just not a decision involving free will.

until Sunday evening or abandon her plan altogether during the week? The answer to that question is complex so Stuart started at the beginning. Once Kate understands decision making she has power to affect hers much more comprehensively.

Enter the dorsolateral prefrontal cortex (DLPFC). Since this may be the first time you've met formally here are a few things you should know about this part of your brain:

- It is a part of the prefrontal cortex (the back, side bit of it).
- Planning takes place here.
- Decision making also takes place here.
- Dopamine (that neurotransmitter with lots of functions) plays an important role here.
- The DLPFC is very well connected to other important brain areas.
- It's involved in working memory.

- It requires lots of energy.
- It gets exhausted quickly.

As is often the historical way, we learn a lot about the importance and roles of parts of the body by observing what goes wrong with them. Decades ago we used to think that it would be a good plan to surgically separate the prefrontal cortex from other parts of the brain!

SLICING AND DICING

In 1935 a neurologist called Egas Moniz pioneered a new operation called a prefrontal leucotomy. He received half the Nobel Prize for Physiology or Medicine in 1949 for this operation and it was used until the mid-1950s.

The operation involved cutting the connections between the prefrontal cortex and the anterior (front) part of the frontal lobe in the brain. According to an article in 1948 in a postgraduate medical journal the operation had successfully treated intractable pain, epilepsy and homosexuality!

The operation also was said to produce results in people with affective disorders, schizophrenia, chronic anxiety and some other serious disorders. Around 10 per cent of people suffered seizures afterwards. Here's where it gets really interesting for us, a dictionary of psychiatry (1970 edition) says 'Post-operative blunting of the personality, apathy, and irresponsibility are the rule rather than the exception. Other side effects include distractibility, childishness, facetiousness, lack of tact or discipline and post-operative incontinence'.

Basically the damage done to the white matter in the frontal lobe altered emotion and feeling, often dramatically reducing them. The patients also became less creative and decisive than before. This shows us how important the prefrontal cortex and frontal lobe are in our daily functioning.

Today we still see patients with prefrontal damage who have trouble with decision making and experience flat emotion and feelings. The prefrontal cortices are so ideal for reasoning and deciding because they are directly connected to every avenue of motor and chemical response in the brain. They can even send signals to your autonomic nervous system, which can produce chemical responses associated with emotion. We'll come back to this shortly.

Keeping our prefrontal cortices in good condition is therefore very important. Unfortunately, the familiar basics are most important:

- Exercise regularly (taking heart-pumping exercise).
- Sleep deeply.
- Take regular breaks.

Filtering incoming data

As you know, we are exposed to a huge amount of data every minute of the day. This information needs to be filtered in order to make any sense of it and to work out what we need to make decisions about and what we can just let pass us by. The anterior cingulated gyrus (ACG) holds the position of chief filterer. It looks at the incoming data, prioritizes and throws out what we don't need.

Subsequently, whenever we are concentrating on something there is increased activity here. It has been noted that people with schizophrenia tend to have smaller ACGs, which may link to theories of overstimulation causing many of their symptoms. If they aren't well equipped to filter out information this could be a component of their illness.

You can help your ACG by having intentions. Thinking about what you want to be aware of and what you want to focus on can help prime the ACG. Kate is very sensitive to people's feelings and changes in their state. She can tell very quickly if someone is uncomfortable in a situation. This is because she is very attuned to incoming data that makes it through her filter. However, ask her the difference between a blue Volkswagen Phaeton and a blue Rolls Royce Phantom after they've both just driven past and she'll say 'um, they're both blue cars'. The differences are filtered effectively for her; they are filtered out.

Everyone's filters are different. Where most people go wrong is by expecting other people to be able to pick up on exactly what they can pick up on. Your genetic programming, your life experiences, your values and desires all have an impact on your filters and they will be constantly changing throughout the rest of your life.

Emotional area

The insular is another very important area of your brain involved in decision making. It is heavily involved in emotional experiences, which are translated into conscious feelings, providing valuable information to help you make a decision. The insular is also aware of all your physical responses, the feelings in the pit of your stomach, it helps you decide to eat when hungry, is involved in your social interactions, and lust and guilt.

The amygdala is the final brain area we'll look at now. It is a big player in the emotional region of the brain and is involved in you feeling uncertain, afraid, frustrated and a range of other feelings. Together these brain areas take in and process lots more information than you are consciously aware of. Consequently, you may sometimes get feelings about things but not be sure why. Try to explore these feelings by tuning into your mind, being quiet and still, and seeing if your unconscious brings anything to your attention. You may find that you suddenly remember something that is very important to the decision you are about to make.

Chemicals

Dopamine is a very important neurotransmitter (messenger) involved in decision making. It is linked to your reward system, your attention system and therefore your decision-making process.

DOPAMINE DECISIONS

University College London (UCL) is responsible for a couple of interesting experiments investigating decision making. The first involved using imaging techniques to detect a signal in the brain that was linked to how much someone enjoyed an experience. It was then found that the signal could predict the choices a person made.

The researchers suspected that this signal was dopamine so they set up an experiment to see what happened if they tampered with the dopamine system. A group of participants were given a list of 80 holiday destinations from all over the world to rate from one to six. They were given a sugar pill and asked to imagine themselves in half of the destinations (so individually in 40 of the original destinations).

The participants were then given L-dopa, which has the effect of increasing dopamine in the brain and is used typically to help Parkinson's disease sufferers. They were then asked to imagine themselves in the other half of the destinations. They rated all the destinations again, then the next day had to choose where they'd go out of paired lists of holidays. The increased dopamine made people choose the holidays they imagined when more was flowing through their brains, so they chose more from the second list.

The leader of the experiment, Dr Tali Sharot, said 'Our results indicate that when we consider alternative options when making real-life decisions, dopamine has a role in signalling the expected pleasure from those possible future events. We then use that signal to make our choices.'

So, higher levels of dopamine make us more likely to rate something favourably and subsequently choose it. Just how far does this effect extend?

INSTANT GRATIFICATION

Another experiment carried out at UCL, this time led by Professor Ray Dolan, showed that increased dopamine makes us more likely to go for something with instant gratification rather than a more beneficial, but slower to be delivered, reward.

Participants were tested after being given a placebo, and also after being given L-dopa (the drug that increases dopamine levels in the brain). The test involved making several choices between 'smaller, sooner' or 'larger, later'. For example, they could choose to receive £15 in two weeks or £57 in six months.

Dr Alex Pine, who was involved in the study, said 'Every day we are faced with decisions that offer either instant gratification or longer-term, but more significant reward. Do you buy your new iPhone today or wait six months until the price comes down? Do you diet or eat that delicious-looking cake? Do you get out your books to study for a future exam or watch some more television?'

The results of Dolan's experiment showed that each of the participants chose more of the 'smaller, sooner' options when they had more dopamine in their brains. This was great for the experimenters as they were only paying out the £15 in two weeks rather than the £57!

So why is this important for Kate to understand? Sticking to an exercise routine is often about the longer-term benefits. Kate must do some preliminary work to make it easy for her repeatedly to choose exercise over other things that come up that offer 'smaller, sooner' rewards. The same principles work for making healthy food choices, doing jobs that you'd rather put off and anything else with a 'smaller, sooner' versus a 'larger, later' scenario.

Currently Kate thinks of exercise as something that:

- she enjoys when she gets to the gym;
- will help her look good in three months' time;
- will protect her from illness in the future (for example heart disease);
- she is supposed to do.

In her mind this isn't setting her up to make it easy to choose it over going for cake with a friend or collapsing on the sofa with a good book. Sensory temptations momentarily boost dopamine levels in our brains, which make us behave more impulsively. If Kate is on the way home from work and sees an advert where a person is relaxing on a sofa with a friend in front of a warm fire with a glass of wine the anticipation of smelling that fire, tasting that wine and being engulfed by that sofa will be enough to trigger the 'reward anticipation' in her brain and release dopamine. We know that it will then be more difficult for her to resist the temptation to call her friend and invite her over for post-work drinks and a catch up (the gym becoming the necessary sacrifice).

Things could be different though. Imagine feeling the same way about the gym as relaxing with friends and wine (or something else you know triggers a dopamine boost for you, for example watching sport or playing computer games). When you reprogramme your-self to create a dopamine boost just from thinking about the gym you are making it much easier for you to follow through with your decision to go. For Kate, who loves measuring progress, even get-ting an activity tracking device may make a huge difference.

Action

Kate's action list

Kate needs to:

- Go to the gym and do everything she can to enjoy it (play her favourite music and tell herself how great this is making her feel).

- Connect to the sensations of stronger muscles while doing weight repetitions, feeling her legs moving quicker on the treadmill, the tightness of her abdominal muscles after her crunches. (This connects her to the benefits each and every time she goes, rather than just in three months' time.)

- Practise thinking about the gym while getting herself into an excited and interested state.

- Anchor the state and the thought together by listening to an upbeat song or spraying a special gym deodorant (that she keeps only for using just before she drives to the gym).

This type of strategy can be applied to anything. Our brain doesn't intrinsically believe it should derive more pleasure from wine than the gym; evolutionarily in fact it should be the other way around. We have conditioned it to respond as it does now; you can choose to start reconditioning it if you want to manage your time more easily.

MYBW top tips for making good decisions

- Keep a clean tidy area to work in.
- Take regular breaks to rest your prefrontal cortex.
- Trust your unconscious to give you feelings and practise tuning into what your brain may be trying to communicate to you at those times.
- Look at what decisions you regularly make and regret (drinking too much on a week night, skipping the gym, etc).
- Work out why you are not making the decision in the moment that you plan to in advance and take steps to make it easier for you to choose that decision.

MYBW top benefits for making good decisions

- You achieve the goals you set yourself because you do the things you need to achieve them.
- You feel great about yourself because you are doing what you say you will do.
- Other people look to you as someone with integrity, this improves your relationships and your ability to influence and lead others.
- Life becomes easier because you make good decisions effortlessly.

5

Is a busy brain a clever brain?

How you learn new things and how to optimize that process

Jessie had new stuff to learn. In the medical world she got used to pretty much always having exams coming up. Ever since she could remember she had been studying. Things were different now from when she was at medical school though, she ran her own business and had to learn on the job, and the learning wasn't straight forward. She needed to get her head around setting up a whole new arm to the business. It had a new legal structure that she wasn't familiar with and would mean working with new clients.

She was concerned about how she was going to remember the things she needed to learn when she wasn't directly using the information every day. She explained to Stuart that she was older now and it was more difficult to learn the older you were.

Recently Jessie went on a training course to explain the ins and outs of community interest companies (CICs) and she was disappointed that she couldn't remember more of it after the training had finished. A secret worry of hers is that if she learns more stuff that it will push the old stuff out of her head. She feels as if her head is pretty full.

There always seemed to be new things that it would be good if she were aware of. The internet was full of inspiring other companies and individuals to learn from, the problem was finding time to fit in learning about them. The other evening she found herself still up at 1.30 am looking at a great organization's website. The flip side of the coin is that she often finds herself in boring meetings barely able to keep her eyes open, much less retain the information that is being discussed.

Situation

Stuart reflected back the key things he thought Jessie was concerned about:

- She is struggling to learn all the new things that were important to her business.
- She is worrying that she won't be able to retain the information.
- He also noted that she was spending a lot of time on work-related tasks.

This chapter looks at how to optimize the way you learn new things and enjoy understanding why this actually means having a balanced pleasurable life.

The brain in action

Your brain is incredible. At a micro level it is made up of millions of cells that communicate to one another. They are communicating all the time, even when you are asleep. These communications are important when we are thinking about how you learn and how you remember.

Your brain makes patterns. These patterns, or organized units of information, can be called schemas. They are created by neurons communicating to one another and forming what is almost like a pathway. This specific track, or schema, can connect with another schema to make a new schema. For example, when Jessie is learning about a new legal structure that she hasn't come across before she takes an old schema and adds the new information to make a new schema. She knew the characteristics of other structures and what the consequences of them were. This is familiar to her – her schema of it is strong. Adding a new characteristic to this schema to make a new schema is easier and more efficient for the brain than starting from scratch.

The brain loves being efficient so any chance it can take to make things easy it will. Stuart knows it is important for Jessie to trust herself with her ability to learn in this new way; her success depends on it. He knows that if she is more in tune with what is actually happening every time she is trying to learn something that she'll feel more empowered by the process.

THE STORY BEHIND HEBBIAN LEARNING

In the 1940s a Canadian neuropsychologist called Donald Hebb came up with a theory for learning and memory. It is based on synaptic transmissions and is now known by many names including Hebbian learning. You'll remember that the way brain cells communicate is by the end of one cell (synapse) sending a message to the end of another cell (synapse). The ends are called axons and they don't touch, there is a gap between them, imaginatively called the synaptic gap. What Hebb discovered in essence is that 'cells that fire together, wire together'.

Cells fire to communicate with one another. Right this moment you have thousands of cells firing and exciting thousands of other cells. Some of these have communicated many times before and so the process is quicker, easier and more efficient between them.

Hebb's discovery is useful to us in lots of ways. For Jessie it is important for her to know that using her new knowledge, either recalling it into her mind or using it practically, she is wiring in the information more strongly. The more she can make a set of brain cells fire the stronger they will wire. Strong wiring means she can most easily recall the information again in the future. It's important Jessie learns how flexible her brain is and how much it is on her side.

In one ear, out the other

One thing Jessie really struggles with is trying to remember things she has heard at training courses. The one she went on recently went at such a fast pace she didn't even get to make many notes. By the end of the day she couldn't even remember what they had covered that morning. Legal and financial things she needed to be aware of just went out of her head and she was feeling it had been a waste of time.

Memory is a complex thing. There are several types of memory and they work in different ways. The type of memory Jessie wanted to use when she was at this training course was her semantic memory. This means that she wanted to be able to recall things that were more concept based rather than experience based. She couldn't relate to the legal or financial concepts experientially, which made them harder to remember it.

It is extremely difficult to remember purely conceptual things over a long period of time after just one encounter with them. For example, if a phone number was given to you by someone you'd just met, or even an e-mail address, you would only have one sense (auditory if you kept repeating it to yourself) to help you commit it to memory. Although this may work long enough to write it down, if you were asked later that same day to recall the number or e-mail address you'd struggle. This is because there haven't been strong enough neural networks embedded.

What makes things stick?

If you are able to look at some key new concepts and create some sort of experience with them, such as imagining a scenario where they played out or explaining them to someone else, then you create emotion and meaning and you create a deeper neural network. If you are able to revisit this on a regular basis (once a day, once a week, then once a month – 'spaced learning') you are able to make the neural network a more heavily grooved one. This means it will be easier to tap into and recall the memory and information when you need it. The best tip of all is to practice recall. Rather than re-reading something, practice recalling everything you can about the topic. Props can make this process easier.

Full head

One of Jessie's fears is that her head is too full. She has been learning things for years and she is a little worried that by learning new things she will push the old things out of her brain. Stuart thinks Jessie may like to learn from a woman who received the Women in Neuroscience Lifetime Achievement Award. When people practise something new Leslie Ungerleider discovered that new neurons get recruited into the neuronal network responsible for that new skill. The experiment was done with people learning finger movements. At first only a few neurons were recruited for the process but with practice more neurons came on board to carry out the task. What was really interesting here was that even without further training the changes in the brain could still be detected. For us this means that when we learn something new at first only a small number of neurons are supporting us to remember it. Over time, with repetition, more neurons come on board to help out. Rest assured we have plenty of neurons ready and willing to become part of a good strong team.

Jessie starts to recognize that there is a very natural process at work here and that if she isn't using information from five years ago then it won't be at the forefront of her memory because it doesn't need to be. If she needs the information again then she will be able to re-learn it and it will be easier than before as there will still be some neuronal network there. She feels reassured by this. In fact she realizes that this is what happens every time she goes back to India to visit family. Her very rusty Punjabi suddenly needs to be resurrected. Each time she worries if she'll remember enough to understand people and tries to brush up on the plane but she invariably finds that more comes back to her than she realized she knew.

Enriching your environment

After her last coaching session Jessie had committed to start doing one new thing for herself every month, starting with a yoga class. Stuart wanted to ensure that this was a habit that got built on over the coming months. He also wanted her to be looking at the bigger picture of how to set her up to be the best learning machine possible for the rest of her life.

In Chapter 1 we saw what Kate learnt from Bill Greenough about the effect of your environment on efficiency and effectiveness in overcoming feeling overwhelmed long term. You'll remember Bill did some experiments with rats giving some 'the rat equivalent of Disneyland' for them to live in. He found that these rats had 25 per cent more synapses than the rats who lived in relative poverty and isolation.

LEARNING FROM RATS; LIVING LONGER AND BEING SLIMMER

There have been many experiments of this type done over the years and the results are all very similar. Three different environments are set up for three different rats. The first environment is solitary confinement, no other rats to play with, limited stimulation, little food and water. The second environment homed two other rats and a running wheel. The third environment homed some of the primary rat's siblings and offspring and a host of toys. They each lived in these environments for several months, after which their brains were examined.

The third rat's brain was significantly larger, had many more neurons than the other two and had lots of neurotransmitters. This rat also lived longer and had less body fat. Here is where it becomes directly applicable to learning. This rat even had more dendrite spines which are the docking points to which other nerve cells connect.

This means he had a larger brain, more neurons, more potential connections and subsequently a greater propensity for learning. The more experiences you have the more access points you are laying down neurologically for new memories to become interconnected with.

Jessie is actually making herself a better learner by experiencing more things in life. If she were to stay in an office from 8.00 am until 8.00 pm six days a week and then sit at home on the seventh day reading gossip magazines, that wouldn't constitute an enriched environment. The consequence would be that she wouldn't find it as easy to learn new things as someone who leads a more rich and varied life.

Knowledge into practice

Stuart knows that Jessie will be best equipped to deal with whatever challenging situation she faces if she understands what is really going on at that time. She needs to learn new things every week and at times she gets frustrated because she knows she's seen something before but can't quite remember what it is. Sometimes she feels that the information has just fallen out of her head!

REMEMBERING

When neurons meet at a synaptic junction and are repeatedly triggered at the same time, on repeated occasions (either by learning something new or by experiencing something) both the cells and the synaptic gap are changed chemically. In future the change means that when one neuron fires the other is more strongly triggered to fire too. The effect is as if these two neurons become partners and when something triggers one they will tend to both fire.

Similarly, if I were to say 'reindeer' to you, you'd probably think of Rudolph the red-nosed reindeer, rather than the Finnish delicacy. Or if I were to say the names Ben and Jerry to you, you'd probably think of ice cream. Depending on how familiar you are with this ice cream you may even get a picture in your mind of 'phish food', hear the crunch as you bite into the chocolate fish and feel the tension in the spoon as you try to scoop up a little more marshmallow and caramel. (Clearly, I'm a little too familiar.)

The chemical change that takes place in the nerve cells and synapses is called long-term potentiation. The neural nets develop a long-term relationship, which we tend to think of as hardwiring. It is important to remember that while this is the way that we can easily remember things and it is how we learn new things, it isn't hardwiring. It is possible to unwire and rewire even after nets have been 'glued together' for decades.

Jessie's grandmother used to say, 'A leopard can't change its spots' so this was always firmly rooted in Jessie. One of her teachers once said to her 'You have to work hard to achieve things Jessie' and this played in her mind over and over again. She knows now that if she ever is thinking about achieving something, unconsciously thoughts of working hard appear.

Working hard is often a good idea and something Jessie often enjoys. Occasionally though she finds herself beating herself up because she knows she hasn't worked hard on something. Examples might be an initial consultation with a new client that she hasn't fully prepared for because she has been so busy organizing the other things she is responsible for; her fitness regime, which is often left to the huge amount of walking she does getting between meetings, but which she often feels is cheating a little and if she was really working hard she'd add a weights regime twice a week. There are always new people to meet and new friends to make. At the end of a long day she doesn't always feel like going out and networking meaning she'd get home with just enough energy to throw herself into bed.

As Jessie talks this through she realizes the type of things she says to herself when she isn't working hard. 'I'm lazy', 'I don't deserve to achieve if I don't put the work in', 'I'll never make it'. She hadn't realized she was so harsh on herself at times. Looking at that underpinning belief that she'd taken on from a teacher about working hard to achieve things had several impacts on her. Some of them she didn't like.

A leopard can change its spots

Thoughts can change in an instant. Jessie used to be a big fan of all things Christmassy, and loved Rudolph and films with reindeers in until one day, as a student doctor, Jessie was involved in the treatment of a middle-aged man. He had sustained a small injury to his nose some time back, but didn't think much of it. He ended up on Jessie's ward. She hadn't dealt with him personally yet so just knew of him as the friendly man in the corner with a Rudolph-like nose (it was enlarged and red).

One day, having moved to another ward, she returned to see one of the nurses on the ward with Rudolph man and she was shocked by what she saw. Most of his nose was missing. The nurse explained to her that he had cellulitis that was resisting all drug treatment and so they had to amputate in the hope of saving his life. In that moment Rudolph changed for her.

Behaviour can also change in an instant. One of Jessie's first patients was a woman called Glenda, who aged 24 had skin cancer. Jessie was taking some blood and liked to distract her patients when she did this so asked her if she had any nice plans for a holiday. Glenda replied that she'd be going back to the Maldives because she'd had such a great time last time. It was an awkward moment for Jessie because she had a duty of care to this woman to ensure that she was educated about the dangers of exposure to the sun for someone who had suffered skin cancer, but she also wanted to keep Glenda relaxed as she was taking her blood. Any sudden movements could have been messy.

Jessie gently asked how she felt about being in the sunshine these days. Glenda exclaimed in a shocked tone that she didn't sunbathe. Ever since she found out she had skin cancer, even though she was successfully treated, she hasn't gone out without sun block on her face and covered her skin. It was no different in the Maldives. She wore a full body wetsuit-type contraption, proudly showing Jessie a picture she carried in her wallet, hat, and stayed out of the sun most of the time. A normally fashion conscious woman had completely changed her behaviour to favour something more important to her.

A lot of hard work

Stuart wanted Jessie to realize just how amazing her brain is. It is flexible; it will change, adapt and grow, as you need it to. He knew she had considered the possibility that she was too old to learn new things now and that anything she did learn she'd just forget quickly. He wanted to show her that even in the most difficult of situations it is possible to learn things.

THE BODY USING THE BRAIN TO DO WHAT WE THOUGHT WAS IMPOSSIBLE

A woman called Nicole had a stroke. One side of her body became paralysed. It used to be thought that if no movement had returned within two weeks that no movement would ever return. The parts of her brain that were previously responsible for movement and feeling on that side of her body had died.

Dr Edward Taub helped patients recover. He would force the patients to use their damaged arms. Using very small steps he shaped their behaviour by encouraging new neurons in the brain to take over the roles of the neurons that had died in the stroke.

By the end of two weeks of practice Nicole had gone from believing she would never be able to do buttons again, to being able to button and unbutton a lab coat quickly. She says 'Your whole mindset can shift about what you are able to do.'

A big part of Taub's strategy came from his early work with monkeys. He learnt that if you just tried to 'condition' a monkey, say by offering them a reward to reach for food, they made no progress. It was only when he 'shaped' them by moulding behaviour in very small steps, being given a reward for effort and small reaches towards the food that they eventually succeeded.

Jessie is realizing that if neurons can learn something as huge as to take on the job of other neurons then learning how to grow a business shouldn't be too much for her brain.

LEARNING CULTURE

A Head of Learning & Development I really respect shared with me some of her experiences in moving to a US-based global company. We had previously worked together exploring neuroscience for L&D and her passion was very clear. When she got to the company something that really struck her was that learning was seen as remedial. They would go on a training course if something was wrong. They would get an executive coach if they were underperforming. At least those were the perspectives.

People being apathetic towards learning is fairly common. But this?

With the help of her team they transformed things. Their focus was that learning is the responsibility of the individual rather than the organization. People were to be empowered to take care of their career. They started to line things up so learning was rewarded and the individual led their growth. Talent reviews would discuss learning. The first step was to create a learning team and function that the team themselves believed in. This then evolved as they created the curriculum, structure and processes which also reflected this mindset.

The increase in engagement with opportunities provided by the organization was huge. While they are still making improvements and learning themselves, this fundamental shift, along with a really strong curriculum was pivotal.

Practically possible

Often we limit ourselves by what we believe is possible. There is a huge body of research alerting us to the impact of what we believe from the thorough and well grounded to the fanciful and weak.

Sayings such as 'You can do anything you put your mind to' can feel empty and lack a sense of grounding in reality.

Getting to grips with things that people have really done widens our true beliefs of what is possible. Below is an example of someone who, with deliberate attention and practice, was able to become an expert in remembering things.

DELIBERATE PRACTICE

Dr K Anders Ericsson is a Swedish psychologist who studies people who excel. He is a fascinating man and has given a lot to the field. One experiment he did was with a normal, average student. Known as 'SF' this student initially was asked to remember some digits and then recite them. He could manage about seven, which is completely normal.

SF was an avid cross-country runner and the number 358 to him meant a very fast mile time – 3 minutes 58 seconds is just short of the four-minute mile. When a number sequence started with a 3 SF coded the rest as seconds and milliseconds. For example, 3493 became 3 minutes 49.3 seconds. After 230 hours of practice in the laboratory he could recite 79 digits. This was better than anyone before, even people with 'photographic memories'. He was able to perform on memory tests to a standard previously seen only by people with lifelong training.

Ericsson could have chosen to assist the average student in excelling at anything; on this occasion he just chose remembering figures. He said from his research with finalists of the 2006 National Spelling Bee 'The main finding is that the amount of solitary study is by far the best predictor of success.'

Jessie was starting to wonder if she'd ever have time for sleep again, and made a comment along those lines. Stuart knew that Jessie liked to see the value behind everything she did, and that she often neglected sleep in order to 'get more done'. So he decided to show her just how important sleep is for learning.

Matthew Walker's research on sleep gives us strong encouragement to use sleep as part of a learning strategy. He found that sleeping between practice or study sessions results in improving on what you have learnt. Sleep restores, refreshes and enhances the brain circuits you need for learning. There is evidence that you even 'practise' what you have just been learning. In 2017 he published a book called *Why We Sleep*. There are now many good books on sleep available.

VALUABLE DREAMING

Pierre Maquet from the University of Liege monitored the brain activity of men playing a video game. In this game they had to explore a virtual town. It lit up their hippocampus (a very important brain area for spatial navigation). Later that day when they went to sleep it was seen that the same areas of their brain lit up.

Another sleep experiment along the same lines involved people who had been playing Tetris. That night they reported dreaming of falling shapes.

Matthew Walker also found that the more a person learns during the day the greater the amount of reply during the night. Walker studied video game players. He found that 'how strongly the hippocampus came back online at night predicted how much better the players would be the next day'. He said 'The more the brain learns, the more it demands from sleep at night.'

So, if Jessie wants to make the most of what she is learning, she needs to get a good night's sleep. Actually listening to her body is new for Jessie.

NAPPING DELIGHTS

The most generic piece of advice we can offer organizations who want people to perform well is to provide places for them to nap.

Google provide 'EnergyPods' and David Radcliffe – the Vice President of real estate and workplace services said 'No workplace is complete without a nap pod'[1] and the neuroscience would agree. Arianna Huffington is very open about the negative effects sleep-deprivation had on her encouraged napping pods at Huffington Post. A more surprising organization embracing the science (because of the typical culture) is the Washington based international law firm White & Case. Other places, like NASA, we would expect to have this under control.

My father used to run a law firm and his colleagues knew that if his door was closed around lunchtime he'd be lying down on the floor by his desk having a power nap. That was 20 years ago!

MYBW top tips for learning

- Set an intention before a training course starts. Have a list of questions you want to get answers to, perhaps some key things you want to walk away knowing.

- Write up key points and refer back to them on a regular basis until you can recall them easily.

- Practise recalling new things you've learnt at random times, such as when waiting for a train. Run through things in your head and later look up anything you couldn't remember.

- Set learning goals with clear outcomes. Set aside time to focus purely on learning.

- Experiment with sleeping for eight hours.

MYBW benefits for mastering learning

- Get more out of learning opportunities; applied knowledge is power to you.

- Stay ahead of competitors by knowing important things that they don't.

- Make more of a difference in your world by being better equipped.

- Do all of this while having a healthy, balanced life that doesn't require you being tied to a desk for 16 hours a day.

6
Getting results easily and with less effort

How to master productive habits and programme yourself to achieve what is important to you

Ben has been working on his action list from last session. Small straightforward things he finds easy to implement. The bigger picture is that he is still snapping at Jane, and her productivity is still too low for this kind of competitive environment. He tells Stuart that she needs to pull her socks up or she'll be out.

Stuart notices Ben is sounding more tired than usual and asks if he is. Ben explains that he and Rebecca argued again last night about him always seeming to be at work and when he wasn't at work they were doing things that he had organized but hadn't asked her about. Ben felt he couldn't win, he thought she'd appreciate him arranging things, it wasn't that she didn't enjoy the stuff they did, it was as if she just wanted to be involved.

Stuart checked how Ben felt he was taking care of himself at the moment while he was feeling these challenges and pressures. Ben revealed that it wasn't very well. He was drinking lots of coffee during the day to try to wake him up and filling up on chocolate, as it was easy and gave him a quick energy boost. He felt embarrassed about this as he was gaining a bit of weight and he used to be really fit and toned. He thought he should just be able to stop, but that seemed to prove harder than he imagined. Overall, he is a bit scared that things seem to be going in the wrong direction. He wants to get control over everything. He knows the type of person he wants to be, it just seems to be more difficult these days. There are many things he needs to pay attention to and things that seem to go wrong before he even realizes it.

Situation

Stuart summarizes the habits Ben has recognized and wants to change. They are:

- snapping at Jane;
- eating sugary food when stressed;
- drinking coffee when in need of energy;
- making plans with Rebecca without asking her;
- working later than he wants to.

This chapter is about mastering what really makes a difference to you achieving results more easily with less effort. It is about the neuroscience behind habits, and subsequently enjoying the extra brain space and energy you have for challenging tasks. The bonus is increased efficiency, productivity and more time for relaxation and other fun things!

The neuroscience behind habits

Some people think habits are genetically programmed and that we have no choice but to follow them. Stuart is keen to show Ben how you can take control of any habit.

Habits are, in their most distilled form, neurological circuits. Just as Ben learnt in Chapter 3 about how synapses communicate to one another to form connections, the same thing happens with habits. The same is true in Chapter 5 in relation to Jessie learning new things. Every time we learn new information we make a new synaptic connection. When we have a new experience we make a new synaptic connection. When we repeatedly engage those circuits we experience 'a habit'. We can take things a level deeper, but be warned… we get into some brain terms!

The basal ganglia plays a key role in the step-by-step formation of habits. The basal ganglia is connected to the area of the brain involved with decision making (the forebrain) and the area that controls movement (the midbrain). It links thought to movement. The main area from the basal ganglia involved with habits is called the striatum.

The striatum receives input from neurons that contain dopamine. They help form habits by providing rewarding feedback to a person in a given situation. This means that the action in a situation becomes easier to repeat next time. For example, if Jane is asking Ben lots of questions when he wants to get on, and he snaps at her, she goes away and Ben can get on with his work. This provides an unconscious hit of dopamine, which makes it easier to respond in that way next time Jane is asking Ben lots of questions.

It is thought that the 'chunking' of tasks is an important component in how they become habits. The repetition of tasks in a series over time becomes a habit. Changing a task in the sequence can disrupt a habit being formed. Similarly, stopping the initiation of the first task in a sequence of a habit can prevent the follow through of a habit.

The reality of the chocolate habit

Some habits are created with the intention of protecting ourselves. Take the situation of Ben reaching for chocolate when he is stressed. Eating chocolate can cause the release of endorphins (which have the reputation of 'happy hormones'). Endorphins reduce pain and decrease stress. If Ben is ever feeling rejected, for example if he doesn't do well on a project or if Rebecca is angry with him, then eating chocolate may make him feel better.

Eating chocolate also leads to an increase in phenylethylamine. This has the effect on the body similar to a very mild 'speed' releasing feelings of excitement and alertness. So when Ben feels he needs a 'hit' of energy or focus he may reach for the chocolate. Other compounds in chocolate activate the dopamine producing receptors, which can also make you feel good, even helping you to focus. But because they're linked to the reward network – they will likely deepen the habit! Another compound, anandamide, is quite similar in structure to a chemical found in marijuana. In order to experience the same high as from marijuana though, you'd need to eat 25 pounds of chocolate. Even with the hours Ben is working, this isn't really practical.

One of the reasons people reach for chocolate when they feel they need a boost is because of theobromine, which acts as a stimulant in a similar way to caffeine in coffee. So it is obvious why Ben is getting some benefit from eating chocolate. However, he is also getting some negative effects (weight gain, feeling out of control, potential tooth decay!). He also finds that the effects soon wear off and the sugar nosedive isn't very pleasant. It is useful to recognize that often habits are created unconsciously with a positive intention.

Ben was trying to stop himself feeling down, rejected, stressed, lethargic and blurry headed. The process of identifying why habits were created can be useful because in creating new habits we ideally want them to deliver the same benefits, or better, with none of the costs.

Once you have decided something needs changing the question becomes is it possible to reprogramme a habit.

Reprogramming a habit

There are now many books on this topic available, most of which focus on the psychological approach to habit change. There are several components to reprogramming a habit. One thing to be aware of upfront is that since 'cells that fire together, wire together' it makes sense to use that. An existing neurological circuit is far stronger than one that has only been used a couple of times. To jump from eating chocolate when feeling you need a pick-me-up to looking at a comical sketch would be a big stretch if you've never experienced the physical effects of looking at something funny before. So the first thing to do when looking to reprogramme a habit is to familiarize yourself with what you are replacing it with. For Ben and the chocolate that could be as simple as looking at a comic for two minutes once a day. (Comical things that are novel trigger the release of dopamine, which would make Ben feel more interested in what he is doing.)

Dr Jeffrey Schwartz is now famous for, among other things, helping people with obsessive compulsive disorder (OCD) to reprogramme themselves. OCD involves very strong neurologically ingrained habits. People with OCD feel a huge amount of emotional distress if, for example, they are forced to leave a room before they have turned the light switch on and off 27 times; others with OCD feel extremely uncomfortable if they haven't been able to wash their hands within the last 15 minutes. Their habits are at the level of compulsion. Compared to this, Ben's habit of snapping at Jane is mild, although for him he feels it is instinctive.

We're going to look in depth at what Schwartz has discovered and how this enables his patients to free themselves from the unwanted compulsions. You will be able to scale this back and use some principles to change any unwanted habits.

OBSESSIONS

From brain scans we know that three areas of the brain are primarily involved in obsessions. The first part is located behind our eyes on the underneath of the brain and is called the orbital frontal cortex. Apparently the more obsessive a person is the more active this region is. It is here that mistakes are registered. Once a person has a sense of a mistake a signal is sent to the cingulated gyrus, which is found in the deepest part of the cortex. Previously you met the anterior part of the cingulated gyrus as the chief filterer.

The cingulated gyrus creates that feeling of dread by signalling to the heart and gut. By this point the person will have a strong desire to do something to get rid of the feeling. The caudate nucleus normally enables our thoughts to flow effortlessly from one on to the next. With OCD though it gets 'sticky'; thoughts are not able to flow. All three of these brain areas are overactive in people with OCD.

Perhaps you have experienced feeling stuck around thoughts or behaviours. Ben finds that when he is working near Jane he notices her slow pace on everything. It's as if he becomes obsessed with her every small move and he feels the pressure rising up until he just bursts out with something snappy at her. He feels he can't help it.

How a faulty circuit can be fixed

The biggest problem with OCD is that the caudate nucleus is stuck; it isn't allowing new thoughts to flow and to subsequently dissipate the negative feelings. Schwartz proved something that didn't start out as a favourable concept. He proved that it was possible, manually, to move the caudate on. He showed people how to overcome their 'stuckness' and free themselves up to do something other than what their compulsion previously dictated. In a nutshell he developed the four-step method, which involves: 1) relabelling; 2) reattributing; 3) refocusing; 4) revaluing.

Patients were encouraged to perform these steps whenever they felt stuck and the results have been thousands of people living free lives. This was a huge advancement. In the past, people thought when you were stuck that was it. We will look at what this means for Ben shortly.

Through Schwartz's methods new neurological pathways can be created which trigger the release of dopamine (because they are enjoyable), which has a positive feedback mechanism that rewards the new activity. This consolidates the whole circuit. Over time, and repetition, this new circuit becomes stronger and therefore gets the competitive advantage over the old one.

Tourettes is very similar neurologically to OCD. Patients with Tourettes report a vague discomfort, an irresistible urge to jerk or vocalize a profanity. Unfortunately, the more they suppress the urge the greater it becomes. Once they have given into the urge they feel relief. The compulsion they feel is similar to the one OCD sufferers feel.

Both disorders involve an inhibition of the circuit that links the cortex and the basal ganglia. The basal ganglia are very important in switching from one behaviour to another. If they weren't working properly it would make sense for compulsions and tics to be present.

TOURETTE'S SUPPRESSION

What happens when people with Tourettes try to suppress their symptoms? Brad Peterson, Jim Lockman and a group at Yale found out by putting Tourette's sufferers in an fMRI scanner (these scanners enable us to see which areas of the brain are active by changes in blood flow). The patients allowed their tics to be expressed for 40 seconds and then had to suppress them for 40 seconds. When the patients were suppressing their tics activity levels changed in the following brain areas:

- prefrontal cortex;
- anterior cingulated gyrus;
- basal ganglia;
- thalamus.

This is the same circuit that becomes activated in OCD and habit formation.

The strategy for helping people with OCD is very similar to the strategy for helping people with Tourettes. From this we can start to think about how best to reprogramme an undesirable habit.

What is possible

TRYING TO MAKE THE PARALYSED MOVE AGAIN

At a hospital in New York researchers worked to try to make people who were paralysed able to move again. This was a huge feat to attempt. The subjects of the experiment had been paralysed by strokes. The first thing to happen was to educate the subjects about what may be possible for them. The researchers started a process which involved wiring new information into the subjects' brains causing their neural circuits to organize themselves into patterns. Next subjects watched a brainwave monitor so they could see how their brains were responding to their thoughts. They focused on moving their healthy limb (strokes normally paralyse only one side of the body) while watching the brain wave pattern. They kept doing this connecting with the unconscious process that their brain was undertaking to make their healthy limb move.

They repeated this to the point where they could intentionally decide to move their healthy limb (without actually moving it). Eventually they learnt to apply those intentional decisions to their paralysed limb. The result was that their paralysed limb could move. It is believed that an important component was the visual feedback they received on the screen by being able to see their brain in real time. They were able to see if they were getting it right or needed to tweak their thoughts in any way.

This is why it is vital to understand what is possible:

- Many problems people experience are because they limit themselves.
- People often feel their challenges are too big to overcome.
- Limiting beliefs prevent progress.
- The power of the mind is huge; often we create what we believe to be true.
- Change is easier when your perspective is in proportion.

Neuroplasticity and habits

Until around 30 years ago people thought that the brain was pretty unchangeable. A person's personality was their personality for life. A person's downfalls were stuck with the individual forever. Now we know that the brain can change, and even that instead of brain cells only dying off from a certain age, that it is possible for them to regenerate. Neurogenesis, the process that generates new brain cells, is natural and very useful.

Just to give you some idea of how much the tides have turned in a very short space of time, let's look at the brief history of neuroplasticity's rise to the top. Initially, Nobel laureates ridiculed Mike Merzenich (a pioneer in the field) for saying that he had shown that the brain cells could change. Merzenich says 'If you had taken a poll of neuroscientists in the early 1990s, I bet only 10–15% would have said that neuroplasticity exists in the adult. Even by the middle of the decade the split would have been 50–50.' Now there is no question that it exists and that it is wise to use it when helping people to get the most out of their life.

THE LOPSIDED VIOLIN PLAYER

A friend of mine plays the violin. She is very good. She played at our wedding as I walked down the aisle. If we were to scan her brain, as many other violinists have had theirs scanned, we would see something really interesting. An area called the somatosensory cortex, which is responsible for our sense of touch, would be enlarged. It was only enlarged for the side responsible for the left hand's fingers though. This is because the left hand controls the strings while the right hand controls the less sensorial important bow.

Statistically she will also have around 25 per cent more of her auditory cortex dedicated to music processing when compared with a non-musician.

What does this all mean? Your brain is changing and adapting every single day. What you choose continuously to do strengthens those areas of your brain making it easier and more favourable to do them again. It makes sense evolutionarily to make you as efficient as possible. Even if you've not been playing an instrument for years, studies show that even by practising some simple finger movements over a few days the relevant areas of your brain will increase.

This means that if you choose to do things that you aren't actually happy with day in and day out then your brain is getting stronger and better at doing those things. You may be snapping at your kids, drinking too much, rushing work that deserves more of your attention or simply telling yourself that you aren't reaching your potential. Waiting to change things can feel as if it makes sense sometimes, for example if you are going through a really busy time and you don't want more things to be thinking about. It is important to be aware that your brain is registering these repeated thoughts and habits. It is getting better at them while you are wishing you weren't doing them.

'Damaged' as a child

Some people feel that if a person goes through difficult times as a child that, because it is such a crucial stage for development, the person will always suffer from the programming that happens at that time. For example, a child whose parents are busy working long hours and who has to compete for their attention learns that bad behaviour gets their attention and time. This continues into the teenage years, where the individual's behaviour gets explained by his or her childhood experiences. While this is obviously a challenging situation, based on brain science, it is possible for the child to change his or her behaviour.

Blaming experiences of the past for current challenges doesn't move people forward; it is disempowering. Compassionately recognizing that the past has an effect but there is hope, and scientific research showing people can change is empowering.

Severe epilepsy can sometimes be treated by removing parts of the brain. In some cases an entire brain hemisphere is removed. There are well-documented cases where children have had this done and they have recovered well. Their remaining brain hemisphere compensates with only minor physical or mental impairments.

If children can recover from having half of their brain removed, it makes sense that they can overcome challenging programming from childhood. The world is filled with many people who have done just this and serve as great examples of what is possible.

THE BOY WHO IS RECOVERING

When I deliver keynotes to organizations, normally afterwards people come and ask me personal questions that they chose not to voice in front of the whole group. On one occasion a lovely lawyer came up to me and explained about the early years of her son's life. She spoke with such composure, and yet seemed to hold such concern for him. I'm sure many people can relate to her fears.

When her son was young he had numerous health complications. This involved lots of hospital trips and stays. It also involved some quite scary, traumatic and forceful medical procedures. When he was home, she would have to continue these unpleasant procedures. The mother was acutely aware that in many ways this didn't feel right. To be pinning her son down (often with help from neighbours) to do something to him he didn't want done. However, it was vital for his survival. She did absolutely the right thing by doing these very hard procedures.

I was personally relieved when she shared that now, aged six, he was responding to hospitals far more healthily. When he was younger he would refuse to even enter them and it would be horrendous trying to even get in the building. Now he happily skips in. When paediatricians tell us that children are resilient, they so often are.

Neural Darwinism

Ben is used to competition. He had to compete to get into the cricket team at school. He had to compete to not be in the bottom 10 per cent of his class each week in tests. He had to compete to get his place at university. He had to compete to get onto the graduate scheme at the accountancy firm. He competes for promotions. He even had to compete to win the heart of his wife.

Stuart uses Ben's familiarity with competition to explain the process that is constantly happening in his head. Gerald Edelman, the US Nobel Prize winning biologist, coined the phrase 'neural Darwinism'.

The concept is simple; the synapses in Ben's brain compete. If they lose, they lose power and those connections become very weak. If they continue to win, they get stronger.

This has been happening to Ben since he was born. In the womb he was producing new neurons at a rate of 250,000 a minute. When he was born the process of apoptosis (programmed mass suicide) begins. The neurons compete to survive; they must show they are part of a useful circuit to survive. Their communication with other neurons with neurotransmitters keeps them alive.

Competitive plasticity

Ben asks Stuart how often he will need to choose consciously to do things differently before new habits are formed and he can unconsciously do things again. This question is a question about competitive plasticity. How much do I need to engage a neural circuit to stop the neural assets being lost to other circuits? Bad habits are breakable and breaking them requires making the new habit more competitively advantageous than the old.

There are many popular self-help advocates of changing a habit in 30, 60 or 90 days. The mere fact that we are told it could take one, two or three months suggests that the science underpinning it is shaky. Personally, I think there are too many factors involved to know for sure how long a habit will take to be a certain strength. Someone could stop smoking the day they find out they are pregnant and never smoke again. Equally, someone can stop eating meat for 90 days and be desperate for a hamburger at the end of that time. Time is only one factor. It can be useful to set trial periods, and by telling people that their new habit will be formed in, say, 30 days you gain some positive priming effects – potentially strengthening those people's self-belief that their new habit is getting stronger each day. However, you can't know for sure what is going on inside each person's mind so to rely on time rather than a comprehensive understanding is less than ideal. It can in fact be detrimental when the habit doesn't stick.

How you create new habits

There are various snazzy methodologies out there for this. The aim is for you to become an expert on you. In many ways you are an individual, motivated by things specific to you. Using generic strategies to motivate you is very hit and miss.

The fundamental process that is required to create new habits, based on what we know from neuroscience, is the formation of new strong neural networks in your brain. This is the same for everyone.

There are a couple of ways of doing this depending on what new habit you want to create.

The first is piggy-backing off an existing strong neural circuit. This would be a great strategy if the new habit you wanted to create were logically linked to an existing one, for example if Ben wanted to work on his habit of leaving the office later than he wanted to. Stuart would be asking him if there are any situations where he leaves the office exactly when he plans to. Ben would realize that whenever he has an appointment somewhere he always leaves exactly on schedule. This habit already has a strong neural circuit, which he could link into in order easily to create the new habit of leaving the office on time. He could try writing in his diary the time of his 'appointment' at home (just as he does for any other appointments out of the office) and then leave enough time to get home for that appointment. This would be creating a whole new neural circuit. It takes discipline and advanced decision making. In order for Ben to increase his energy when he is feeling low he currently drinks coffee. To create a new habit here he will likely need to start from scratch. To help make it as easy as possible Ben needs to consider what already gives him energy (apart from coffee). Does he ever close his eyes for five minutes and then feel refreshed? Does breathing deeply for a minute help? Does listening to a blast of an upbeat song for 30 seconds do the trick? Whatever already works for him, or he can imagine working for him, is the thing to use to create the new habit.

This first step doesn't create a new habit, though it is vital to the next stage. Once you know which method of creating a new circuit you are going to use you are ready for the second step. For this you need to work out or learn how to get yourself to act out the new habit repeatedly. The circuit will become strengthened with repetition ('cells that fire together, wire together'). Your personal motivation is something that you need to become an expert on by trial and error. This isn't something that you can just follow a three-step plan for. If there were a three-step plan claiming to tap into your personal motivation then it would be hit and miss if it would work for you!

Stuart cannot tell Ben what he wants to do instead of drinking coffee. Nor can he tell him why it is fundamentally important to him to start doing it. He can offer suggestions of why other people may think it is important, but what Ben takes from that is up to Ben. Finally, Stuart cannot make Ben choose to do that new thing for the first time, the second time or the number of times **along with** the mental attitude that will create the new habit for Ben. Only Ben can do that.

Internal intrinsic motivation is often the most powerful way of quickly creating a new habit. If you can derive enjoyment from the actual new habit you are creating it will serve you in the long run. There are some groups of people in the world who have great intrinsic motivation and subsequently appear to have great discipline.

Directing new habits

Awareness is very important with habits. Being clear about what you ideally want to create, do and think shapes you. Willed mental activity actually changes the structure of the brain. That is why it is so important to bring Ben's habits into his conscious awareness so he can choose what he wants his life to be made up of. The process of coaching is actually facilitating self-directed neuroplasticity. Ben is choosing how he wants to reprogramme his brain and Stuart is helping him. Ben puts in the hard work of directed mental force. Stuart helps him see himself from different angles.

Protecting new habits

Once you have put the effort into recognizing old habits and decided on the most sensible new habits to reach all your outcomes, you want to protect them to ensure that you can rely on them.

Your brain can be likened to a magnet and iron filings. Imagine a desk with iron filings on one side and a magnet on the other. When you run a weak electric current through a magnet, nothing happens. When you run a strong current through it all the iron fillings come rushing over to it. A similar thing is happening in your brain. When your synaptic connections are weak, they don't have much attraction power. When they are strong, they attract new neurons to support the circuit further. Neurons will automatically be drawn to where there exists electrochemical activity.

This means that the more you can light up a new circuit the stronger it will become. The brain doesn't distinguish between real or imagined when you are lighting up circuits. Ben can imagine feeling tired at his desk and this triggering him to walk downstairs, step outside his office, breath deeply, feel refreshed and return to his office. This mental rehearsal will activate the neural circuit and help attract new neurons to it and make it easier to 'choose' that good new habit until it is stronger than the old habit.

Part of the process involved in continuously being able to learn and strengthen new circuits is to weaken old circuits. Evidence shows that decreasing the neuronal activity in circuits will lead to weakening of them. Practically, this means not paying old habits attention. Any time a thought of an old habit comes into Ben's head it would be great for him to use that as a springboard to jump on to another thought. Focusing on an old habit, running it through your head, even while saying to yourself 'I don't do that any more' gives strength to the circuit and so is a bad idea.

Why bad habits sometimes return

PAVLOV'S RATS

It is believed that emotional memories aren't removed completely from a person. Rather, it is more like they are prevented from expressing themselves. Pavlov's work with rats backs up this idea and gives us an understanding of why old habits sometimes return.

Pavlov found that over time previously extinguished responses could spontaneously recover. In one set of experiments he found that if a rat was conditioned with a tone and electric shock in one box and then moved into another box the fear response could be 'removed'. If the rat were put back into the original box then when it heard the tone again it would experience the fear response again.

This would be like Ben going on holiday for two weeks and not eating any chocolate or drinking any coffee then returning to work and finding his old habit returned.

It was also found that stress could reinstate a previously extinguished response. This means that if Ben changed his chocolate-eating habit, when faced with stress he would be open to it returning. Being aware of this gives Ben the opportunity to put things in place to decrease the chance of him choosing anything other than his new desirable habit.

Action

The experiment with Pavlov's rats highlights why it is so important to keep consciously working on new habits until they feel like second nature. The aim is for your sense of identity to shift to the degree that the old way of doing things doesn't even sound like you any more. Embedding a new habit often requires self-control, so only tackle one or two a month to maximize your chances of success.

If you are going through a particularly stressful time then recognizing this is the first step. Next make a plan that will reduce the likelihood of you making undesirable choices. Ben may choose, for example, to take popcorn and dried fruit to work to snack on. He may spend 60 seconds at the start of each day picturing how he wants it to go and what he will do when he feels weak. It might even be that Ben takes a relaxation class or listens to a relaxing audio to reduce his stress before starting work.

Ben's action list

- Write in his diary the time he needs to be home as an appointment, practise thinking of it as an appointment, and subsequently always leave enough time to get home for it.

- When he arrives home on time for 10 consecutive days, experiment with when his energy levels are dropping at work going outside for five minutes and taking his iPod to listen to upbeat music while breathing fresh air.

- When he has found a successful new way of re-energizing himself, look at his habit of snapping at Jane. Set up a meeting to talk to her about it.

MYBW top tips for habits

- Realize that you can change almost anything.
- Give new habits lots of energy through practising them or imagining doing them.

- Work on only one or two at a time that require self-control.
- Use existing strong neural networks to set up new habits where possible.

MYBW top benefits for mastering habits

- Experience increased efficiency, effectiveness and productivity.
- Gain a sense of accomplishment knowing that your life is examined and on track.
- Gain more cognitive processing space and energy to focus on what is important to move you forward.

Part 2
YOUR COLLEAGUES AND CLIENTS

Very few professionals work exclusively alone. Most actually have to interact with colleagues and or clients on a daily basis. Understanding how your brain works is great because at a fundamental level, that's how your colleagues and client's brains work too! Trouble can set in though when we make that mean that people will behave in the same way as we would in a given situation. Here Stuart works with Kate, Jessie and Ben on challenges that involve other people in some way. Being able to work effectively with others can make life much easier. We know most people leave jobs because of relational challenges.

7
Working and living in balance

How to design and create your world to work optimally

Kate is exhausted. She is naturally a bright, bubbly person and even when she is running on reserve energy supplies you can see her real personality shine through. Stuart knows things are taking a toll on Kate though. Not being a whiney person, Kate didn't mention at first how tired she felt. She doesn't even allow herself to step back and look at everything to see what the balance of things were.

Kate feels guilty even talking about work–life balance because she organized the training session that happened at work on this very topic. It's never been a concept that resonates with her though. She hears the term being bandied around, and knows people say it is essential but it sounds to her like something weaker people need.

The training at work covered all the general points such as getting enough sleep, having holidays (where you turn the smartphone off), taking time for meals, spending quality time with friends and family and subsequently being sprightly for work. The general consensus though was that the trainers had flown in from outer space and had no grasp of what it was like to work in a corporate environment. Their suggestions and guidelines showed no appreciation of the real world with all the expectations that come from both work and home.

There seems to be a lot of hype around being able to 'switch off' which Kate resents. The idea of sitting in front of a television watching mindless 'fly on the wall' documentaries seems like a waste of her time. Even being told how she should be 'unwinding' drives her mad.

Recently, at work she feels her boss is demanding more from her but she doesn't have faith that if she succumbs to these extra demands they will benefit her in the long run. This is leaving her feeling more stressed and in conflict about what to do.

Situation

Stuart recognizes Kate is uncomfortable even talking about this topic so looks at a more generalized level of abstraction when checking he has understood the situation. This approach helps Kate to feel better about addressing things that she was nervous about. He suggests that based on what she has said they look at how she knows she is creating her optimal week.

This chapter looks at designing and creating your life in balance, whatever that means to you rather than what others may dictate to you. The bonus is experiencing less conflicts with others.

Work–life balance

In general usage work–life balance relates to the balance between work and personal time. Most people think of it in terms of the division of time between the two. For example, a person who leaves for work at 7.00 am and returns only at 7.00 pm, and then has his or her phone on to answer questions or e-mails from 8.30 pm until 10.00 pm would typically be thought of as having poor work–life balance.

The history of working hours is very interesting. At times, people have been encouraged to believe that more leisure time will make you happy. At other times, more belongings have been cited as the things to make you feel fulfilled.

There is no single answer to the question of what makes a good work-life balance. Essentially, people have programmed themselves over time to respond to different situations and want different things.

One thing is for sure though – good organizations are aware that requiring people to work all their waking hours is not wise.

Currently, more women are working outside the home than ever before and one of the biggest causes of stress in women is balancing work and home lives.

WARMTH AND PARENTHOOD

In one study by Cuddy, Fiske and Glick (2004) several interesting points are made:

- When working women become mothers they 'trade perceived competence for perceived warmth'.

- This doesn't happen for men when they become fathers – they keep their perceived competence and additionally gain perceived warmth.

- Scarily, people 'report less interest in hiring, promoting and educating working moms' when compared to childless women or working fathers.

These kinds of views and realities have an impact on working mothers. Kate has felt that in order to remain a valuable member of the senior team at work she needs to put in long hours there. She also feels that as her girls are getting older in a lot of ways they need her less, but in some ways she feels it is important for her to be around more. So she does feel torn.

Let's take the metaphysical definition of balance; a desirable point between two opposing forces. When we consider work–life balance this presupposes that work and life are in opposition from one another! Unfortunately, this is some people's reality. If that is the case then it is best to rectify it as soon as possible.

So is there a perfect balance? Are there a number of hours one should spend on work and another number for everything else that would give balance and a happy fulfilled life? Absolutely not; it cannot be the case. The effect of one-hour's work in a state of flow, for example experienced by a professional sports person, could be completely different to an hour experienced by a frustrated lawyer. Equally, a fulfilled lawyer in flow could emerge more energized from a 10-hour workday than an unfulfilled juggler who works 8-hour shifts with the circus.

Work and life are two terms that vary from person to person. The consequences of them are variable and, subsequently, to create a balance between them is unique to individuals. What constitutes 'life' for an individual is very important. In the past, for Kate it has meant rushing around from one commitment to another. While she enjoyed being there for people, most of her activities were very similar (listening to people and trying to help them). A really easy thing she can do to increase her balance would be to mix up what she does with friends. Inviting them to a yoga class where they can both exercise their body and mind in a different way, or going to learn a new language together once a week, would activate areas of their brains not routinely used in that way.

It may be that for one of Karen's team members they would love to leave the office early on a Wednesday so they could go and play a round of golf with their father. That simple thing would enrich their whole 'work-life balance' perception.

Expectations: theirs and yours

So who gets to decide upon and create your work–life balance? Many people feel that their employers have a big say in this. Some people complain that their company expects too much from them, or that people outside work have unrealistic expectations about their work commitments. Communication and education here are key.

Ideally, before accepting a job somewhere you will have a good idea of the expectations of the company, your boss and your team of the normal, and abnormal, levels of time and energy they expect you to put in. You can also get a good idea of how they feel about lives outside work – are these respected and encouraged as good for networking, developing you as a person and making you an overall better employee? Or not so much? If these things weren't clearly understood before you joined you could create a training opportunity to explore them now. Looking at why the brain works best with a variety of experiences, rather than just work ones, can provide a good starting point to opening the dialogue of what would best serve the company.

REALISTIC RECRUITMENT?

Some organizations are wising up to a disparity between what is 'sold' at recruitment fairs verses what people experience when they start work. In order to attract good talent some places have found themselves extolling the opportunities of 'flexible working' and 'remote working'. This is attractive to many people. However, sometimes they are not fully aware that those ways of working may not be available to them in their role. Actually a lot of travel will be involved, or periods of very long hours. These are major lifestyle impactors.

Mis-selling reality is a really bad plan because the negative feelings that build up towards the organization are big. This toxicity is contagious and can be leaked to those who may have originally been quite happy with the work-life balance culture.

Kate needs to become a designer and then a detective. First, she needs to design her life the way she'd love it to be covering things such as:

- when she wants to start work and finish work on a typical day;
- what type of tasks she wants to do at work;
- how many holidays she wants to take a year;
- what she wants to do when she isn't working;
- how often she wants to see how many friends;
- what time and what activities are spent with her family;
- what skills she wants to develop outside work.

These are examples of basic ingredients that Kate needs conscious awareness of. From there she needs to become a detective to discover how everything can best fit together.

The detective, curiously and methodically, is responsible for fitting things together. This can take a period of trial and error to gain feedback data that is valuable on better orchestrating things. Initially, you can start with your best guess and create a type of template of a month's ideal activities. When you have feedback you can go back to that template and tweak things to increase their usability. The other important role for the detective is in working out how to communicate this most effectively to all the people involved in your life. You should ensure that work colleagues understand how you will be working and why and what the benefit of that is to them. You need to enable your family and friends to be clear on how you make decisions about how much you see them. This is all down to you patiently explaining things.

Conflict

Conflict can arise when people don't understand your plan. Even more conflict tends to arise if **you** don't understand your plan, most commonly because there isn't one. Imagine the situation where you have eight good female friends, each of whom are now married. When these eight friends were single you used to see them on a regular basis, once a month each (which meant seeing a friend two nights a week). You'd also see them in a group environment a couple of times a month. Once they got into relationships the expectation would covertly change into seeing the couple, but then also seeing the friend on her own too.

Before you know it, you're in a relationship too. Now the invites are coming in for double dinner dates (so the partners can meet each other and then all get to know each other better) and girls' nights. You also have the date nights with your new partner. Each of your friends has the unconscious expectation that things won't change now you're all in relationships. This means that you go from having two invites or commitments a week to five – and that's with only one date night for you a week. Repeatedly saying no to invites, or not providing as many yourself, can quickly lead to resentment if things aren't openly discussed. So the key in this instance is to ensure that people know they are important to you. Consider other ways of keeping connected other than seeing them at the same frequency as before. We know that people create meaning when things aren't crystal clear (sometimes even when things are). So minimize the risk of people feeling left out if the design of your life is changing by clearly and repeatedly communicating with them.

The other main type of conflict to look out for is your own mental conflict. If you aren't clear on your ideal design of life then it can be easy to feel the conflict. When an opportunity to take on another project comes up, but you had already committed to helping your partner with some home renovations and improving your golf game before the summer. Having 'if-then' plans and lists of priorities for different quarters of the year can make difficult decisions easier.

Causes of tiredness

Tiredness is often cited as one of the indicators that things are out of balance. So what causes it?

TIREDNESS

When you are awake and your brain is active a chemical called adenosine accumulates in an area called the basal forebrain. When the adenosine gets to a certain level it attaches to specific receptors in that area. This has the effect of inhibiting the neurons (preventing firing, which is the method of communication). This means that you start to feel more tired. Interestingly, caffeine primarily works by blocking adenosine, by attaching to the receptors that adenosine usually does.

There are other things going on when you start to feel less alert and more tired. For example, your brain being awake and active requires glucose for energy. As these levels start to drop chain reactions are set in place that also lead to you getting more tired.

All the classic things that you are aware of can zap your energy and leave you tired:

- exercising too much or too little;
- eating too much or too little;
- sleeping too much or too little;

- being mentally active for too long, or not long enough;
- doing the same thing too many times, doing too many different things;
- worrying too much.

Basically you're Goldilocks in disguise!

Time versus energy

Kate often finds that she uses her diary exclusively to tell her whether she can or cannot do something. After last session she understands how decisions are key to her time management, and she is realizing today that they are key for balance too. Previously, if a friend were to text her during the day asking if she wanted to pop round for a cup of tea that evening she would check her diary. If there were the time available to do it (even just a gap between work and Pilates followed by cocktails with another girl friend) she'd say yes. Then she'd just work through her lunch break and take some bits home to finish after cocktails, or get up early to finish off.

This strategy meant that she felt good about herself because she was programmed to have as part of her identity that she was there for her friends whenever she could be. The downside to the strategy was that it didn't take into account a longer-term view of things. Running the strategy for a week would be fine, longer than that could start to have a negative impact on Kate's energy, simply because it doesn't factor into the decision-making process.

Since they had worked on similar concepts last session Stuart was keen for Kate to use what she had learnt already to make some suggestions about how she could do things differently. That way she was engaging her brain to think and would take ownership of her ideas.

Kate suggests:

- Keeping a log of her energy levels in a normal week to see what tends to trigger a decrease in them. (This will involve cataloguing how many social things she attends, with whom, what exercise she does, what foods she eats, and anything major that happens at work.) She realizes that this isn't a scientific experiment but hopes it will give her some pointers.

- Reprogramming herself to value being there for her friends in her optimal state. Practically, this means not just fitting people in when it is likely to tire her out. It isn't a hard and fast rule though, because sometimes fitting people in will energize her – it depends on who and what they do.

- Considering and experimenting with ways to re-energize after spending time with people who do drain her.

The same principle applies to work. If an extra project is put on the table and someone is expected to pick it up Kate will normally do it. She wants her colleagues to feel that she is a team player and hardworking and that she will do her fair share of the work. The problem is that often Kate is working at full capacity and so this extra work pushes her into the category of too much work. Ultimately, this isn't best for her co-workers in the medium or long term because Kate becomes drained.

The importance of control

Our sense of control is one of the most important things to us from a neuroscience perspective. It is logical that we want to be in control of what we do and what happens to us. When we don't feel that we have control over situations we can feel very uncomfortable, a threat response may be generated in our minds and bodies and we can respond protectively. In Chapter 13 we see the neuroscience behind a threat response.

The fact that Kate describes new things that her boss is asking her to do as 'demands' is very telling. It indicates that she feels that a degree of her control is being taken away. This in itself will stress Kate out. Even if the thing being demanded of her fitted into what she wanted to do and she felt that it would be a good career and life move for her, the fact that it comes as a demand is a big problem.

The other problem with these demands is that she isn't sure she wants to do them. If they were presented as requests then she'd feel she had more leeway to decline them. What Kate's next move is depends on the quality of her relationship with her boss and her communication skills. Ideally, she wants to be in a position where she can talk to her boss and work out between them what the best plan is for everyone involved. This type of relationship and level of communication can take time to build. First, she needs to step back from the situation and mentally reframe things to just look at what is on the table. Does she want to do the things or not? For some people they are entrenched in the concept that they have to do things. Kate doesn't **have** to do anything. There will at times be consequences if she doesn't do things, as there would be with almost all jobs. However, she can choose not to do things or to communicate and look for another way to reach everyone's outcome.

While sometimes when we talk about regaining control in your work life it can sound removed from the reality of the situation, that is actually a good thing. When you are in the situation you can feel you have no options and no control. When you look at it from a step back you realize that you do have choices. This is both the reality and the best way of looking at things. At times in life, choices are taken away from us and we can feel as if we have no control over things. Finding small ways to exert control is vital even in these times.

VIKTOR FRANKL ON FREEDOM

Viktor Frankl gave us huge insight into personal freedom in his book *Man's Search for Meaning* (1946). In this he concludes from his experiences as a prisoner of war that the freedom of choice you have in every situation is vital. He is quoted as saying 'Everything can be taken from a man but one thing; the last of the human freedoms – to choose one's attitude in any given set of circumstances, to choose one's own way.'

He also gives us great advice about balance: 'What man actually needs is not a tensionless state but rather the striving and struggling for some goal worthy of him. What he needs is not the discharge of tension at any cost, but the call of a potential meaning waiting to be fulfilled by him.'

Choosing goals in work and in life to move towards that are meaningful to us, challenge us and that we have a chance of achieving gives us a strong foundation to work towards. Difficult balancing decisions are much easier to make when you have a core base of goals you're working towards.

BAD BOSSES

One of the most common challenges to time management and work-life balance is bad bosses. This is quite blunt.
Heaping new urgent work on people is poor management. Somewhere along the line of control something has gone wrong. In the vast majority of organizations it is possible to plan and manage expectations so that sudden short deadline work isn't required. This would remove a HUGE amount of undesired last minute work.

How the mind works optimally

The brain itself needs balance. It likes:

- clear expectations;
- well-articulated goals that are achievable;
- to be challenged;
- rest and relaxation;
- variety.

Creating a life that has all of these things in it takes planning and monitoring.

Action

Checking the design and creation of your life is something that is well worth doing on a monthly basis. That way years don't pass without you consciously engaging. You are in a position to appreciate every month of your life and notice if things start to go off track and are able to take action to correct things. Looking back over a 20-year period saying 'I spent too much time at work, I've missed my kids growing up' or 'I wish I'd have worked smarter and achieved more in my career' can be avoided.

MYBW top tips for work–life balance

- Decide what work–life balance means to you.
- Plan how to make your ideal work–life balance a reality.
- Communicate with the people involved, clearly and repeatedly.
- Check your life regularly.
- Take control of everything, starting with your internal experiences.

MYBW top benefits for mastering work–life balance

- Decisions become easier on a day-to-day basis, enabling you to be more efficient.
- Other people around you have clearer expectations of you, leading to fewer conflicts.
- You set yourself up for knowing that 20 years down the line you are proud of the decisions you've made and the life you've led.

8
Upgrading your life one step at a time

How to get to grips with what your brain needs to help you achieve your goals

Today Stuart had some ideas of what he wanted to cover with Jessie. He checked to see if Jessie has anything pressing on her mind or whether looking at her goals would be a good use of the time.

Jessie often feels that there are lots of things to do. Running her company means that she has overall responsibility. If she hasn't thought about it then she has only herself to blame. If she hasn't made something happen then the buck stops with her. This means that she is often working on many different things at the same time. She doesn't always keep track of them all very efficiently. Sometimes she feels it is potluck whether she remembers something or not.

At the moment it is really important to her to expand. She has so much work coming in that she knows she needs more good people in place to take care of it. Some people in the great team she already has want to take on more responsibility. It's up to her to be organized enough to train them and then give them more responsibility. She knows delegation is a buzzword, but never feels she has enough time to actually do the delegating.

One of her big goals at the moment is to get a social enterprise arm of the business set up. These are called community interest companies (CICs). Becoming a CIC would enable more people to know exactly what that side of the business did and how they could help and benefit from it.

Recently, Jessie was invited to a dinner with some people she didn't know well and some people she didn't know at all. She almost didn't go, but ringing in her head were wise words from an old mentor of hers: 'Say yes more than you say no.' She ended up sitting next to the owner of a very large company who loved the concept of her business. He set up a meeting with her and they are now one of her biggest clients. She just wished she knew how to replicate this!

Situation

Understanding goals properly will enable Jessie to work more easily with her colleagues and clients. She will find she is more fulfilled because she is aligning her expectations and actions overall, giving her more integrity. The specific examples she has raised are:

- keeping track of all the things she is working on;
- developing her team to take on more responsibility;
- strategically move big goals forward;
- replicating opportunities.

This chapter is about achieving your goals by aligning your brain to work with you on all levels and free up thinking resources to be able to focus on other things, while increasing your credibility with others and trust in yourself. The bonus is you can confidently move onto tackling more complex goals.

What the brain has to do with getting stuff done

The power of the mind is huge. There is a lot of varied research that has led people to explore deeper into this power.

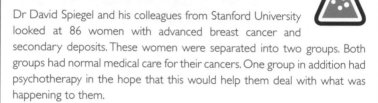

WHEN YOUR MIND COULD BE THE DIFFERENCE BETWEEN LIFE OR DEATH

Dr David Spiegel and his colleagues from Stanford University looked at 86 women with advanced breast cancer and secondary deposits. These women were separated into two groups. Both groups had normal medical care for their cancers. One group in addition had psychotherapy in the hope that this would help them deal with what was happening to them.

The women who received the psychotherapy had an average life span of 37 months after being diagnosed. The women who didn't had only 19 months.

At UCLA School of Medicine Dr Fawzy has done a lot of research into this area. One of his studies involved a group of people with malignant melanoma skin cancer. Half of the group participated in a 90-minute education and support group weekly for 6 weeks. This group of people experienced less fatigue, less depression and better coping abilities than the other group. Six years later three people from this group had died, whereas 10 people from the other group had died.

Engaging your brain in what you want to happen is incredibly important. We used to think that if you wanted something to happen you would just get on with it. If it wasn't in your physical power to make happen then it was out of your control. Some people still believe that. A growing number of people believe that mental power is actually a very powerful force and creates real changes. Stuart decides to introduce Jessie to three experiments before focusing on goals. Some of these experiments may at first seem totally unrelated, but they all demonstrate the power of the mind. Until Jessie is convinced that her brain has a part to play in achieving her goals there is no point in talking about them because she will only ever achieve a fraction of what she is capable of.

EXERCISE AT THE PHYSICAL GYM AND THE MENTAL GYM

Drs Guang Yue and Kelly Cole proved that by imagining you are exercising a muscle you are actually strengthening the muscle. One group did physical exercise and one group imagined doing exercise. The exercise in question was a simple one for the finger.

The routine was a strict one, involving 15 maximal contractions with 20 seconds' rest between each, on Monday to Friday for four weeks. The group imagining the exercise simply imagined doing the exercises, but also imagined hearing a voice shouting at them 'Harder! Harder! Harder!'

The group who put in the physical hard work increased their muscle strength by 20 per cent. The group who did the mental hard work increased their muscle strength by 22 per cent. A few decades ago this would have been seen as impossible.

Now we know that when we give our brain a command, such as 'contract finger muscle', even if we don't physically do it the neurons are activated and the circuits strengthened. When the muscle does eventually contract it is stronger.

Jessie is amazed by this and immediately makes the connection to how she can use it to achieve her goals more easily. By realizing that the thoughts she has are having a powerful effect on her body, even something as simple as what she says to herself at the start of the day becomes important. Recently she is often saying to herself 'I've got so much to do' and she feels her body ever so slightly arch. She can't say she feels motivated, more overwhelmed. She makes the decision right then to ban that phrase from her vocabulary. It may be true, but focusing on it isn't helping her. Instead she decides to say to herself 'I'm enjoying making progress on valuable tasks'. She realizes that these things can all be taught to her team too to help them gain the same benefits she is enjoying.

HEALING WHAT IS NO LONGER THERE

Phantom limbs cause amputees a lot of problems. When a person has a limb amputated the brain still has a neurological map of it – in effect the brain still thinks it exists. Unfortunately, this means a person can still experience pain in the amputated limb. A great man called V S Ramachandran worked out a way to help many of these sufferers.

The process involved placing a mirror in front of the healthy remaining limb, which creates the illusion of a present missing limb. The brain acts as if it is present and the patient can stretch it, exercise it, even itch it. This is now routine treatment for many amputees.

Some people weren't yet able to undergo mirror therapy. G L Moseley, an Australian scientist, worked with people to get them ready for the therapy. Moseley believed that by building up the amputated limb's motor map in the brain it might start to change. He trained the patients to imagine moving the painful limbs. This activated the necessary brain networks. They were also given lots of pictures of hands, right and left, and their aim was to identify which it was (left or right) quickly and accurately to activate their motor cortex. They even practised imagining hands in various positions for 15 minutes three times a day.

After 12 weeks of therapy, including the mirror therapy once they were able to take part, in 50 per cent of people the pain had gone. For others it was reduced. Years ago it would have been seen as impossible to cure someone of pain without surgery, electricity or even medication.

People can achieve amazing things when their brain is optimally programmed. Jessie is starting to get excited about what she may programme her brain to do. Stuart wants to look at one more experiment before they focus on goals.

Being open to opportunities is very important in business, especially in an entrepreneurial-type organization. If Jessie misses opportunities, such as the one at the dinner party, it could set her back months. Seizing opportunities, on the other hand, could mean her business reaches vital tipping points much quicker.

CAT'S EYES

Colin Blakemore and Grant Cooper, when at Cambridge Psychology Laboratory, did some experiments with cats that links to seeing opportunities. Two groups of kittens were placed in separate environments. In one there were horizontal stripes, in the other vertical stripes.

The effect of this was that as the kitten's sensory receptors were developing they were being limited. The 'horizontal' kittens couldn't see vertical objects. When they were put in normal environments, for example with a chair, they'd walk straight into the chair leg, as if it wasn't even there. The 'vertical' kittens would choose not to walk on a tabletop, or would walk straight off it.

You can only 'see' what you have programmed yourself to see. If Jessie wants to avoid missing any opportunities then she needs to be programmed to see them for what they could be. We will look at exactly how to do this a bit later.

Brain areas of goal achievement

Our ability to accomplish goals is closely linked to our ability critically to appraise our own actions and the actions of those around us. These processes rely on our frontal lobe; specifically, the part of the frontal lobe called the prefrontal cortex plays a central role in enabling us to achieve our goals. It is crucial in both us creating goals and then creating the action plans to achieve the goals. The process involves identifying the cognitive skills you'll need to implement the plan, then coordinating those skills and using them in the right order.

Once you have gone through a whole process of planning and enacting your goals your prefrontal cortex evaluates how you did. It makes judgement calls of success or failure when measuring the result up to your initial intentions. Many people in the self-development world say that you shouldn't ever label something as a failure. This may have merit; however, your brain will still be declaring something a 'match' or 'non-match' to your initial intentions. You can absolutely focus your attention on what you can learn from any 'non-match' results and just know that unconsciously your brain registers what has happened.

Getting specific

Stuart decides to start pinning Jessie down on some of the things she'd like to be different in her life. He starts with one of the first things she mentioned today involving feeling on top of things. Jessie had said she didn't like feeling as though there were lots of different things to keep moving forward and it being potluck whether she remembered them all or not.

This could easily be turned around into a goal. Often classic goal setting tells people that goals have to be tangible and so they should avoid any goals related to 'feeling' a different way. This is a limitation on the coach's side, not the brain's. Currently with Jessie feeling as she is, she is setting herself up to feel out of control. This could have negative consequences. For example, when she goes into a meeting and is feeling out of control she is programming her mind with her expectation to be all over the place. She will be engaging her fear-response programming so she will impede her prefrontal cortex thinking clearly.

This doesn't mean that Jessie has to get everything sorted in her life before she can start setting goals or expecting to see major progress towards the realization of the goals. She just needs to start somewhere. Jessie and Stuart decide that the first goal is for Jessie to feel on top of all the different things she is working on at the moment. Most classic goal-setting training will tell you that the goal is currently in poor form. For example it doesn't fit the SMART format (specific, measurable, achievable, realistic and time-targeted). There is merit in having structure to a goal; however, many models aren't based on what we know about how the brain works and so only cover part of the picture. You could have a perfectly formed SMART goal and still not achieve it. Equally, you could have many 'poorly formed' goals and achieve them all.

The question Stuart asks Jessie is 'What would need to happen for you to feel on top of all the different things you are working on?' This gives Jessie an opportunity to break down the goal into its constituent parts that she can then work on. She mentions wanting:

- a list of everything that is current;
- a separate list of what she's working on each day;
- all her clothes to be hung up at home;
- to make a packed lunch each day.

Interestingly, there wasn't as much as Jessie first thought there would be for her to feel on top of things. She was worried she wouldn't be 'allowed' non-work things, such as her clothes and packed lunch. Stuart explained that how we feel isn't always logical or rational to us. If we intuitively think doing something will make us feel better then it is often best to do it. Stuart asks her 'How would you feel getting this all sorted?' and observes that she is connecting with the emotion and feeling of what achieving the goal will give her. She seems to visibly relax and smile.

Real or imagined

Once you are clear on what you are aiming for the next step is to look at how you're going to get there. Being aware of the benefits of rehearsing can be very rewarding. The purpose of rehearsing is to get better at something. You can rehearse different parts of a something. For example, when a saxophonist practises one time they will focus on finger movements and they may drill a certain phrase over and over again, without even blowing air through the instrument. The aim is to get quicker and quicker until the neural web in the player's brain automatically kicks into action when he or she starts that sequence in future.

Breaking a goal down into the constituent parts that need to occur in order for the whole thing to come to fruition is useful. With some goals this is harder than others. Taking her goal to make a packed lunch each day as the example, what could be the constituent parts? Jessie suggests getting up 10 minutes earlier than normal, sitting down and planning things to go in packed lunches before she does the shopping, actually making the lunch, washing up the lunch box each night and taking time to eat it at lunch time.

Jessie realizes that any one of these components could trip her up and that looking at them has opened her eyes. She needs to address these things strategically and mentally. Currently, for example, her brain is programmed for her to get up at a certain time. She needs to reprogramme it to wake up 10 minutes earlier. She then needs to programme it to want to get straight up out of bed, go down to her kitchen and make her packed lunch. This part of her goal alone takes several steps, all of which have the potential to not happen.

Jessie makes an unusual but wise decision. She decides to focus on one component of one goal a day with an aim to embed the over-all new habit within a month. She will take one component and actively strengthen the synapses relating to that component. For example, to start with Jessie focuses on waking up 10 minutes earlier than usual. The next day her focus is on actually getting out of bed. One the third day she focuses on getting out of bed again, but also in the evening checking she has things she can use to make a packed lunch. On the fourth day she focuses on getting out of bed and making the packed lunch during that extra time she has created. On the fifth day she is aiming to make the lunch and also eat it at lunchtime.

STRENGTHENING YOUR CONNECTIONS

Just as driving around a dirt track creates grooves in a certain path, the connections between the synapses in your brain can be thought of as being in grooves. We say that there is an increase in synaptic strength when this happens.

When you want to achieve a goal you may need many things to be working in your favour including:

- positive anchoring;
- good habits;
- good decisions;
- an effective strategy.

Many of these components will be more programmed in your favour if you are consciously strengthening the pathways. Strengthening pathways can be as simple as:

- Thinking about the thing you want to happen. For example, this might be going to the fridge and choosing fruit over chocolate.

- Thinking about the process you will take to get there. For example, to get a toned bottom imagine getting up in the morning and doing 50 squats while cleaning your teeth.

- Actively working on a component of your goal, while keeping the big picture in mind. This might be, for example, tackling a boring project while having in the back of your mind your promotion.

The neuroscientist's biggest kept secret

There are many great discoveries in the neuroscience world that have not yet reached the business world. This is one of the biggest, most profound secrets and has some of the most far-reaching consequences. We're looking at it here with a view to applying it to achieving goals but it has many other applications. It is called priming.

Priming activates certain neural circuits and sets us up to respond in a certain way. Practically, if Jessie tells herself she is a bit scatter-brained when it comes to keeping on top of everything that she is doing, that primes her in a certain way. The pure act of identifying herself as that person activates certain circuits in her brain. You'll remember the priming experiment where people read documents to

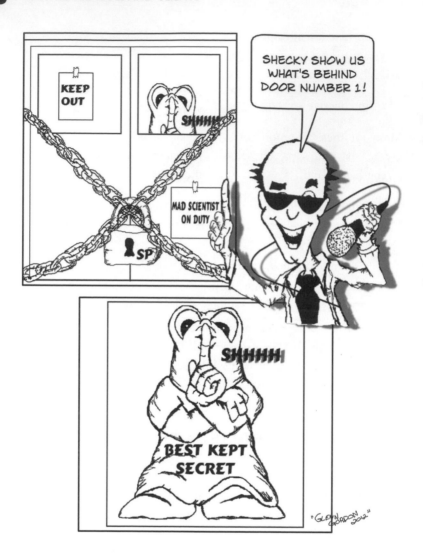

prime them as either a secretary or a professor. They then either performed more quickly (a secretary is associated with being quick) or more accurately (a professor is associated with getting things right). This experiment is very important for the achievement of goals.

Achieving goals depends on several things. From a big picture perspective it is our strategy and our ability. What most people don't realize is just how delicate and easy to influence our ability is. Even 10 years ago people would explain ability to be something that you could or couldn't do. You could improve your ability by practising and training. Only recently has it become more mainstream to consider the effect your mind has on your ability. For many people this is still an alien concept. When you look at the actual science behind it, rather than just getting caught up in the 'touchy feely' world you understand why looking at what your brain is doing is vital.

Thinking about how you can use priming in the achievement of your goals is limited only by your imagination. Working out what type of person, or which actual person, would get the best result in a situation is a good start. Many entrepreneurs say they have role models or people they aspire to be like. This may be as far as they take it. For some they ask themselves 'What would Richard Branson do in this situation?' for others they distil the person's values by reading lots about them and then take on those values as their own.

There are easy ways to prime yourself; anything that activates those neural circuits works. Experiment with different things to see what works for you. Is it simply thinking about a person and how he or she behaves? Is it talking to the person? Is it reading about someone? Is it watching a film or television programme about them or things related to your goal? The more different ways you can activate the circuits the better.

If Jessie set a big goal to be a really fit, healthy, balanced business owner Stuart would set her a few pieces of homework. They might be:

- Find three role models who run very successful businesses and are also fit, healthy and balanced.
- Watch at least one documentary about a person like this.
- Read at least one book on a person like this.
- Upload to your phone an audio you can listen to for five minutes every day.
- Talk to one person who you believe is the next stage on from you in this.
- Visualize as you lie in bed for one minute before you get up each morning this new you, considering the choices that you will make today.

This would help Jessie to activate properly the circuits she'll need to support her to achieve her goal.

Motivation

People often say that they can't get motivated or that low motivation is standing in their way to achieving their goals. It isn't just an excuse (although sometimes people do use it as one). It is definitely uninformed though. The prefrontal cortex, which we've met before, is involved in self-motivation. So in order to get started you will need to engage that part of your brain. Remember your prefrontal cortex uses up lots of energy and needs frequent breaks.

Motivation can be thought of as what initiates, guides and maintains goal-oriented behaviour. When you start working towards a goal, taking small steps of action, you become hopeful. This activates the reward system in your brain and causes the release of dopamine. The dopamine makes you feel good. This makes you want to keep going and to repeat actions that made you feel this way. The whole process of working towards a goal becomes pleasurable. You have activated your brain and body's natural way of keeping you motivated.

Your state matters

FOCUSING ATTENTION

The effects of 9/11 terrorist attacks were far reaching. The first study of the psychological symptoms and post-traumatic stress highlighted some important points we can learn from. This study was carried out by the Research Triangle Institute in North Carolina. They found that there was a distinct correlation between reports of post-traumatic stress and the number of hours a person spent watching the attacks and their aftermath. In short, the more television they watched about 9/11, the more likely they were to suffer psychological effects. This correlation was not affected by whether or not someone had lost a friend or family member in the attacks.

Where your brain is focusing your attention can have a powerful effect on your state and what happens in your life. The study about people watching 9/11 footage illustrates this. Experientially we know that our focus affects what we notice. For example, when Jessie was transitioning from a normal job into working for herself and becoming an entrepreneur all she kept coming across was other entrepreneurs. She'd see books about them, television programmes on them, meet them out and about and find herself thinking about different famous ones.

TELEVISION'S REALITY

Obviously we know that not everything we see on the television is real. We know that actors are hired, that scenes are stages and scripts are performed. But does all of our brain know that?

The right hemisphere processes images from acted programmes the same as from reality programmes. Our limbic system goes one stage further and actually responds as if what we're seeing is really happening. This means that when you watch a horror film where the bad guy is creeping up on the good guy your threat response is being activated.

What state would be most beneficial for you to be in to achieve your goal? Jessie realizes that in order for her to be calm, on top of things and organized she does need to be in a certain state. For her, listening to the news on the way into work may not be the best plan. She feels frustrated, angry, that she can't make a big enough difference, and confused about humanity when she listens to what is happening in the world. Knowing now that her brain is releasing chemicals as a result of this and that in turn she is priming her brain activating certain circuits, she decides to experiment with a change.

Jessie decides that to maximize her self-management she wants to listen to *Delivering Happiness* by Tony Hsieh in the mornings instead of the news. His audio book (2012) is all about his entrepreneurial story and is very inspiring and motivational, as well as containing lots of great business lessons.

Deserving a reward

Jessie never saw the logic behind giving herself a reward as she progressed towards a goal. For her, she either achieved the goal, which was a reward in itself, or she didn't, in which case she didn't deserve a reward. It seemed too fluffy to give herself rewards along the way. The brain actually loves rewards though, releasing good chemicals to help you keep going and want to take similar action again.

Action

Jessie says that she will trial for a month breaking up one goal into much smaller milestones and rewarding herself along the way.

Jessie's action list

- Have a meeting with Sarah and start her sales training. Buy *New Scientist* magazine as a reward.

- Give Emma books and online presentation skills training and arrange a date for her to give a presentation and get feedback. Take mum out for dinner as a reward.

- Organize for an external strategic trainer to come in to work with everyone on strategy for next year. Book a facial as a reward.

MYBW top tips for goals

- Consider both your practical strategic plan and your mental strategic plan.

- Mentally visualize the components required to achieve your goal regularly to strengthen the synaptic connections.

- Start with the goals you really want to achieve and break them down.

- Prime yourself to be the type of person who has achieved such goals.
- Plan rewards for yourself for the achievement of components of the full goal.

MYBW top benefits for mastering goals

- Your brain is working with you, rather than against you, to achieve your goals – making the process easier.
- Dedicate more of your thinking resources to other things once you have strategically set yourself up to achieve your goals.
- Build a new identity being able to trust yourself always to achieve what you set out to achieve.
- Increase the trust others are able to put in you.
- Progress to more complex goals with more variables – achieve these faster.

9
The minefield of motivating people

The vital aspects of motivation that aren't usually mentioned

Ben has been having a great week. He has been getting home at the time he said he'd be home each day and his wife is really happy with that. When he feels an energy slump coming on he has been going outside for some fresh air and a burst of music and this has perked him back up. It wasn't always easy to choose this over the chocolate and coffee, but he is determined and knows only he can make that choice. Overall, he has been feeling as if he is getting on top of his personal management.

This week he decides to talk to Stuart about the other people he works with. Ben really likes the fact that he gets to work in a team because he believes teams are important and that you should be able to get more done by working in one. This week he is really struggling with motivating some of them though.

Jane is an ongoing challenge for him. He knows she should be capable of working more productively but he doesn't really know how to make her. He decided to have that meeting he'd been meaning to have with her. What he found out was actually a lot more insightful than he expected. It was as if flood gates were opened and when someone was actually listening to her everything just poured out.

She was worried about losing her job, scared that he was constantly angry and disappointed with her, frustrated at herself for not being stronger and hadn't got over her boyfriend dumping her. (Although she did say that obviously personal problems were left at home.) Ben wasn't really sure what to do with all this information, because the bottom line was he still needed her to do a good job. However, he wanted to be compassionate and help her to not worry about stuff so much too.

Last week he ran a training day in another office about their internal technical bespoke software. The participants were all polite, but they didn't really seem to engage and get excited about the topic. One of the guys in the team called James seemed to get distracted a lot and Ben felt he needed to work hard to keep him on track. It seemed as if James was more focused on saying things to get a response from the others in the group than on the work. Ben worried he wasn't motivating him in the right way.

Later this week Ben's team will be given another case to work on and he really wants it to be better than ever before – for them to all come together and approach it with some gusto. From Ben's point of view they all have to work together on the projects so they may as well do it to their best ability.

Situation

Stuart knows that motivating other people can have a big impact on many people's quality of life so this session is important both for Ben and his present and future colleagues. They need to look at:

- helping Jane be productive;
- increasing lines of communication between Ben and Jane;
- engaging and motivating people like James;
- team motivation overall.

This chapter helps you understand how motivating others really works, increases your potential to help others be more productive and decreases the frustration for you and your colleagues. The bonus is less stress at work and home.

Motivating others

Motivation is a strange phenomenon. We tend to recognize it most in ourselves or others when it is lacking. It often feels as if it's something we cannot control, something elusive at times and something to be worked to the maximum when it is around.

So what actually is motivation? Motivation shares many similarities with being proactive, which is why we are starting here. Proactivity is about more than taking initiative it is about a core responsibility. The decisions we make are pivotal here. The word responsibility can be changed to 'response-ability'; our ability to choose our response.

Responsibility is a concept that is very familiar to Ben. In the coaching world one of the first things you ensure that people you work with understand is the concept of responsibility. This is so that they understand that the power to change things lies with them. Whingeing about external things consistently doesn't tend to lead to an improvement in a person's circumstances. Of course there are certain external circumstances that are beyond your control.

Victor Frankl grew up with Freudian psychology which stated that whatever happens to you as a child shapes you and sets you on a path that governs the rest of your life. As a Jew he was imprisoned along with his parents, brother, sister and wife in a Nazi death camp. Only his sister and he survived. Suffering torture and atrocities that no human should ever endure, never knowing what his future held Frankl started a journey that has given the rest of us a wealth of insight. He talks in his book (originally published in 1946) about 'the last of the human freedoms'. He explains that the Nazis could control his whole external environment and could do whatever they chose to his body, but they couldn't control his basic identity. How these experiences affected him was a choice he would always control.

We can think of motivation as a precursor to the decision to do something. When a person is motivated enough to get out of bed, they make the decision to get up and then before they know it they are up and getting ready for the day ahead.

What motivates you

We can think of motivation being of two main types, intrinsic or extrinsic. Intrinsic motivation means it comes from within. There are no external forces at work; you do something because you enjoy it. Extrinsic motivation on the other hand comes from outside you. You do something in order to gain something else. Typically this is money, recognition, some other form of reward or the avoidance of punishment. There have been a lot of studies done on intrinsic and extrinsic motivation. The key is to understand when to use which.

Some rewards motivate

WAITING FOR REWARDS

Anticipated rewards are powerful from our brain's perspective. Mathias Pessiglione and his colleagues showed that some areas of the brain, including the ventral striatum and amygdala, affect our behaviour when we anticipate rewards. Other parts of the brain are important for assigning value to different things in different situations.

For monkeys a reward that really motivates them is, for example, a banana or apple. When they see one their orbitofrontal cortex lights up. The cells that light up in this area of the brain seem to discern a type of hierarchy of rewards. The cells fire more for an apple if shown an apple and a piece of lettuce but more for a banana if offered with an apple.

Most of us have had the experience of something external enticing us and making us more likely to do something. Parents (rightly or wrongly) are often seen offering tasty rewards in exchange for good behaviour or the achievement of some task. As adults it isn't uncommon to hear someone motivating their partner to do the washing up with the promise that they can watch the rugby, uninterrupted, afterwards. Experientially we know these kind of external motivations can work on occasions. Soon we will look at what effect they have from a big picture perspective.

TRANSLATING MONEY INTO POWER

Another study carried out by Mathias Pessiglione and colleagues demonstrated the power of unconscious motivation. They showed that people will adapt the amount of effort they put into things depending on the level of reward they expect to gain.

The study involved showing a picture that had been masked to individuals. The picture was either a £1 coin or a 1p piece. They were both masked to the extent that the individual couldn't tell you that it was there. Unconsciously though the people had picked it up. Money was used because it has consistently been shown to activate reward circuits in the brain.

The individuals' brains were monitored, as was their skin conductance and hand-grip force. (The hand-grip force is recognized as a measure of behavioural action.) They were shown a screen where a thermometer illustrated their increasing or decreasing behavioural effort through squeezing. They were told that the higher the thermometer went the more of the money they could keep.

The results showed that the £1 motivated individuals to put more effort in than the 1p, even though they weren't consciously aware that there was a difference. The areas of the brain that were primarily involved are those that are thought of as the output channels for the limbic system. This carries out emotional and motivational activities.

Dopamine is stored abundantly in the nucleus accumbens, which is very sensitive to other neurotransmitters such as serotonin and endorphins. You'll remember that serotonin, endorphins and dopamine make you feel rewarded and satisfied, which really drives motivation.

Perhaps you can relate to this the experience of monkeys with lesions in their nucleus accumbens. They would rather eat a peeled nut now than hoard unpeeled ones for later. Many people suffer from the effects of low motivation for rewards that are coming up in the future. For example:

- sitting on the sofa rather than hitting the gym to sculpt that beach body physique versus all the last-minute things people would prefer to try;
- saving money each month versus using a credit card when big expenses crop up;
- learning and then disciplining yourself to work efficiently during the day versus working late when the office is quiet.

The trick is to become a detective to identify ways to get a dopamine boost while you are doing the repetitive, sometimes boring tasks that will lead to a longer-term pay off and reward. Everyone is different in terms of the best things to trigger a dopamine release but here are some ideas:

- Have a checklist so that you can tick off your achievements. Seeing your progress can be useful.
- Create specific focuses to differentiate each task (for example biceps one workout, triceps the next; labelling cash deposits as different things).
- Listen to music.

REWARDING CHILDREN

This classic motivation experiment involved researchers studying children during their 'free play' time. They noticed the children that consistently chose to draw during this time. This indicated they were intrinsically motivated to draw because they chose to do it when they could do anything.

These children were divided into three groups. The first group would be rewarded for drawing. They were shown a 'good player' certificate and asked if they'd like to draw in order to receive the certificate. The second and third groups were just asked if they'd like to draw. If children from the second group chose to draw then they were given a certificate. This was unexpected though, not part of an 'if… then…' deal. The third group received no reward.

Two weeks later the children were observed again during free play time. The children from groups two and three drew as much as they had before. Those in the first group though, who had been offered a certificate if they drew, showed much less interest now. This is known as the Swayer Effect. What had once been play had been turned into work.

This research is pivotal and was re-analysed and followed up by three scientists. One of the scientists, Edward Deci, said 'Careful consideration of reward effects reported in 128 experiments led to the conclusion that tangible rewards tend to have a substantially negative effect on intrinsic motivation.'

External motivation can be used effectively for very basic repetitive, non-brainy tasks. However, nothing that Ben or his peers are likely to come into contact with falls into this category. The dangers of using it without understanding it are as follows:

- It can reduce or remove a person's intrinsic motivation.
- It can kill creativity.
- It can force people to think short term.
- It can lead to unethical behaviour.

Higher purpose

One of the best ways of connecting with your intrinsic motivation is to know, at the deepest level, why you are doing something. The best way of illustrating the effect this has on individuals is to actually look at a company. In *Built to Last*, Collins and Porras (1994) dedicate a whole chapter to 'More than profits' in which they identified that having a purpose bigger than money was a distinguishing feature of companies that would go the distance.

Masaru Ibuka created Sony in 1945. Within 10 months he had a prospectus for the company that included the following purposes of incorporation:

- To establish a place of work where engineers can feel the joy of technological innovation, be aware of their mission to society, and work to their heart's content.

- To pursue dynamic activities in technology and production for the reconstruction of Japan and the elevation of the nation's culture.

- To apply advanced technology to the life of the general public.

This was before the company was even cash flow positive. It is unusual, but thankfully becoming more common, for companies to be so clear on what, other than to create profits, they exist to do. Since the initial prospectus, the 'Sony Pioneer Spirit' was created, a part of which states 'Through progress, Sony wants to serve the whole world... Sony has a principle of respecting and encouraging one's ability... and always tries to bring out the best in a person.' That's easier to get out of bed for than 'to make lots of money' isn't it?

Individuals need higher purposes too. Why are you here? What are you contributing during your time on earth? What difference do you want to make? These are questions the answers to which should roll off your tongue. Being able to link your daily activities, even the repetitive ones, to your higher purpose helps keep up your intrinsic motivation.

Power of expectations

Expectations are powerful things. They can actually change the incoming data your brain pays attention to. When Ben was working hard to keep James on track he could have used his expectations more. Stuart goes through the basic things Ben needs to know about working with expectations. We can think of an expectation as the brain's way of being aware if there is a reward or threat on the horizon. When an expectation is met, you experience a dose of dopamine and have a reward response. This helps programme you to remember what you did and do it again. If you exceed your expectations, you get a strong hit of dopamine and reward response. Getting it spot on can cause brain functioning changes similar to being given a dose of morphine! However unmet expectations cause a large drop in dopamine and a threat response. You'll remember

that dopamine is great for thinking and learning so meeting and exceeding expectations puts you in a great state to get things done.

Ben could try a few things with James and the rest of the group:

- Set expectations. For example, this might involve asking each of them to write down how many documents they expect to check through that morning or what their outcome from meeting with a client is.

- Set the expectations a little lower than he actually expects the group to achieve (to help them exceed them).

- Frame things so the group sees all the small ways they have exceeded expectations.

Motivation zappers

When Ben found out that Jane had been struggling with some pretty big concerns he wanted to help her for her sake and also for his. When working with other people, as Ben often does, sometimes it is valuable to explore what may be going on for them. Stuart stresses that Ben can only guess what is going on as he is not actually coaching Jane. The thought processes people go through can be useful for Ben to go back to Jane with though. Stuart asked him to outline all the things Jane was worried about:

- losing her job;
- feeling scared that Ben was angry with her;
- feeling scared Ben was disappointed with her;
- being frustrated at herself for not being stronger;
- not having come to terms with the end of her relationship.

This may not be everything that Jane is worried about. However, the fact that she has shared these issues is a good start and a great place for Ben to start with her.

Stuart begins by discussing with Ben the brain background and why dealing with these challenges is so important. Our brains are wired to try to maximize our reward and minimize any danger or threat. We typically move towards reward and away from threat. As we know, what is rewarding and threatening to each of us can be slightly different but we can make some assumptions too. The limbic system (the collection of brain areas that support emotion and behaviour) is very sensitive and so any of these concerns will certainly be enough to arouse it.

The threat response is very powerful and long lasting. Much more so than the 'toward' response you get with a potential reward. The threat response has an impact on your ability to think. It reduces your cognitive resources. You'll become more defensive, be less able to process new information and you're likely to start seeing more things as threats. This is why Jane is less productive than either she or Ben wants her to be.

Once Jane's train of thoughts start going in the direction of worry, and she starts to experience the associated emotions, it is very difficult to stop the slippery slope. Just trying to bury the emotion doesn't work, so being in an environment where you believe you have to hide your emotions can be challenging. Suppression is a very energy-intensive process. Studies show us that it has several effects including making people around you feel uncomfortable.

Instead of suppressing an emotion there are two other things that are useful. One is to label the emotion and the other is to refocus attention elsewhere. So when Jane is feeling scared that Ben is disappointed with her she needs either to take action to find out if that is the case or she needs to refocus on something else. If she has done something wrong then she needs to find out how to rectify it or apologize and look at how to prevent it happening again. Different situations will require a slightly different blend of these components. For example, if she is scared that Ben is angry with her about something but he seems really busy, then the best plan may be to refocus on something else until it is appropriate to talk to him (which may be at the end of the day or in the morning, depending on how Jane knows Ben likes to work).

The other thing that will help Jane is to label the emotion that she is feeling. For some people this is an unusual thought. It has been shown, however, that by labelling how you feel you reduce the emotional response you've been feeling. In order for this to work the label needs to be succinct. A short, symbolic label rather than a long rambling description about how she feels. For example 'frustrated' or 'irritable' or 'sluggish' rather than 'mad at Ben because he isn't being clear enough with me and that makes things really difficult...'. The longer description tends to increase emotional responses and draw a person further into feeling more of those feelings.

All of these strategies take time to master but are worth the investment. Teaching these ways of doing things to other people is best done with their permission and buy-in. So explaining that you've come across some strategies that have a scientific basis for them and which have been proven to help people is a good place to start.

With Ben's new found efficiency, effectiveness and productivity he proudly told Stuart that he would schedule a 15-minute appointment with Jane to 'sort things out'. Stuart realized that Ben's expectations here were possibly unhelpful. He explained that when dealing with other people, especially around potentially (or definitely) delicate issues then how much time you spend on it shouldn't be your focus. If it is your focus then you can be concentrating on the clock rather than the individual. This can lead to you missing important indicators and the whole meeting being less effective. Unfortunately, in a worst case scenario this can do more harm than good, or at best require another meeting to put things right.

Instead Ben needs to allow plenty of time for a meeting of this nature. His expectations should centre more around how he will show up, who he will be for Jane, how he will handle her questions and where he hopes to get to in the meeting. If it takes 20 minutes then he has unexpected extra time he can use to do some odd jobs (of which he should have a list, like a rainy day activity list for children). If it takes 60 minutes then he can feel good knowing he has given Jane quality time.

Power of control

There are certain things that predispose a person to being motivated. One of these is having the sense of autonomy over what they are doing.

Experiments have shown that the important thing here is a sense of autonomy. In many organizations it is difficult for a person to be

ADDICTED RATS

There is a famous experiment which was carried out by Steven Dworkin, who is a professor in psychology at the University of North Carolina. The study involved two rats. The first rat was able to give itself a dose of cocaine by pressing a lever. The rat eventually dies through lack of food and sleep. The second rat is given the same dose of cocaine at the same frequency as the first rat chose to administer. He died much sooner.

100 per cent autonomous, and this isn't necessary to get the benefits. There are small things that can be done to boost a person's sense of autonomy. These things do have to be real though; they cannot just be paying lip service to increasing employees' sense of autonomy. Companies often reap benefits well beyond what they imagined when they do this though. Companies could allow employees to be responsible for:

- Choosing their own tea break time and what they do during it. (Some companies don't even allow this.)

- Discerning what clothes are appropriate for work. For example, if they are seeing clients it is their responsibility to dress up, if not they could choose to dress down.

- Working on something that shows corporate social responsibility (ideally choosing the team that they work with on this and when they meet and how they report on it).

- Making up work time if they want to take time off. (This is far more prevalent in Australia than in the UK for example.)

Some companies give their staff more autonomy. At 3M, some staff are able to spend 15 per cent of their time on whatever they choose. It was during this time that Post-It™ notes were created. Imagine the lost revenue to the company if that had never happened. Google also give engineering teams time to work on whatever they want, for them 20 per cent of their time. Ideas such as google suggest, gmail and Orkut are among the many things that have come from this time. The company says that 50 per cent of new creations come from this time.

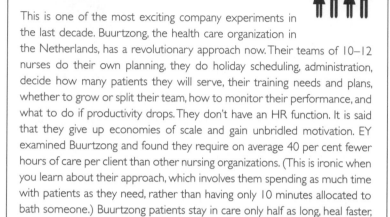

BUURTZONG

This is one of the most exciting company experiments in the last decade. Buurtzong, the health care organization in the Netherlands, has a revolutionary approach now. Their teams of 10–12 nurses do their own planning, they do holiday scheduling, administration, decide how many patients they will serve, their training needs and plans, whether to grow or split their team, how to monitor their performance, and what to do if productivity drops. They don't have an HR function. It is said that they give up economies of scale and gain unbridled motivation. EY examined Buurtzong and found they require on average 40 per cent fewer hours of care per client than other nursing organizations. (This is ironic when you learn about their approach, which involves them spending as much time with patients as they need, rather than having only 10 minutes allocated to bath someone.) Buurtzong patients stay in care only half as long, heal faster, and become more autonomous. The estimated savings for the Dutch social security system if all home care organizations achieved Buurtzong's results are close to €2 billion every year.

The numbers speak volumes. However, they don't tell of the restored joy a nurse gets from doing her job properly, from giving to her community in the way she knows is best. They don't mention the fact that the nurses are excited on Monday mornings and feel fulfilled on Friday afternoons. However, the fact that absenteeism for sickness is 60 per cent lower and staff turnover is 33 per cent lower than traditional nursing organizations does make most people sit up and take notice.

Power of certainty

In Chapter 13 we look at the core components of great leadership from the brain's perspective. We explore what basic needs your brain as a leader has and the brains of those you lead have. Certainty is one of those needs. Stuart raises it with Ben now because in his role as Jane's leader he could make things much easier for himself and for her if he understood the power of certainty.

Going back thousands of years, imagine you live in the great outdoors. You have to hunt for your food and rely on the other people in your tribe to secure your survival. If you fall out with people in the group you will be outcast and this will almost certainly spell death. Not knowing when the next meal is coming is a threatening situation because if it doesn't come soon people will die. Being uncertain about which berries are safe and dangerous is potentially life-threatening. Uncertainty threatens survival.

Our brains still respond in a way that strives to restore certainty. When Jane isn't sure what Ben is thinking she feels uncomfortable. When she is worried that he is thinking bad things about her she gets caught up in it, as it is a major threat to her. She fears she will lose her job, be unable to get another one, not be able to pay her bills and have to move in with her mother. From Ben's perspective he sees her not getting on with her work at an acceptable rate.

Restoring Jane's certainty would really help her to let go of the unhelpful thoughts. Ben could do this by being more open, clear and communicative about things. Of the initial challenges she shared with him three of the four could be alleviated in this way. If Jane understood what specifically Ben was angry about and disappointed about and what she could do to ensure that didn't happen again then she would feel more in control. Equally, if Ben made clear that everyone makes mistakes and he will always let her know if the mistake is very serious and may lead to disciplinary action, which may eventually lead to her losing her job, then she can stop constantly worrying about job security.

The other challenge Jane raised, that she hadn't got over the end of the relationship with her boyfriend, isn't solved by Ben's more open communication. This is a separate issue, and it may be best for Ben to elicit from Jane what she thinks she should do about this (for example talk to a girl friend or a personal coach). Stuart is keen that Ben doesn't think he has to solve all of the problems of the people he works with and can distinguish between things that he can easily help with and those someone else may be better placed to assist with.

Power of confidence

Confidence is another of the core competencies that a leader needs to be aware of how to generate within the people he or she works with. One of the ways confidence can seem to be increased or decreased is with the raising or lowering of a person's status. One of Ben's team, James, finds this a particular challenge and, unknown to both Ben and James, diverts a lot of his focus onto trying to raise his status in the group.

So why is status so powerful? Evolutionarily higher status was directly linked to higher rewards. Some would say that's still the case today, others would say not. Either way the brain is still very definitely programmed to activate the reward neural circuits when perceived status goes up. Equally, the brain experiences a threat response when status goes down. We can see why this would be so – a lowering of status could lead to all sorts of negative consequences.

For many people, something as simple as talking to a boss can activate the threat response. Most people perceive their boss to have a higher status than them and so, as a consequence, their status is lower. It is all in our minds though and if the effects of the threat response are causing problems, working on shifting your perception can make a big difference.

MEETING GRIFF

Being quite narrow in my television watching habits I don't come across that many television presenters. At a Birmingham Civic Society Leaders' Breakfast there was a friendly looking guy called Griff as one of our guest speakers. I had understood that he was the president for Civic Voice. After thoroughly enjoying his funny, illuminating and motivational talk I was suitably impressed with him. I knew he was going on to the Civic Voice talk that I was going to and so I asked if he knew the way, which he didn't, but he offered for me to walk with him as he was walking with a gentleman who did know the way. Wanting to know more about this charming great speaker I asked if he did lots of these talks all over the country. He replied that he did some. Curious I asked what he got up to when he wasn't doing this. The pause alerted me to something being amiss. He regained his pace quickly and reeled off a wonderful list of things that he was involved in, including television presenting.

The kind gentleman we were with chipped in saying that Griff was indeed involved in lots of great things. It wasn't until we arrived at the next meeting that I sensed people behaving a little unusually towards him. Their body language was different around him. Their tone of voice was altered when talking about him and to him. I considered that perhaps they'd all heard him talk before and were impressed. During the break I phoned my husband and mentioned that we'd had a great talk that morning from the president of Civic Voice, Griff Rhys Jones.

After being shown several television programmes he has been in I feel different about meeting him again. Not only is he now a great live inspiring presenter in my eyes, but he is also that guy from the television!

Bosses and famous people are obvious ways people tend to feel lower on the status scale. There isn't a fixed list of ways people can feel that their status is being lowered; here are some more common examples though:

- seeing someone wearing the same shirt as you, but thinking that it looks better on him or her;
- hearing that a friend has bought a new expensive car;
- watching someone else's child win a sports day race, while your child comes second;
- hearing your colleague be praised by your boss.

When people's status gets lowered they often try to raise it. This is usually an unconscious process. Many comical sketches revolve around this situation playing out. Imagine the conversation by the water cooler where one person mentions how they are getting a new car. The other person just happens to mention that their little Billy got an A in chemistry and his teachers think he will go to Oxford. A third person pipes up that they are volunteering with the homeless this Christmas.

On the surface they look like just a group of people sharing some news. Some would even say that the sharing is bonding them together. Unfortunately though, when people are trying to increase their relative status they are decreasing their bonds with the other people. This means that when James sits in meetings and comes out with things that are trying to distinguish him and raise him above the others he ends up isolating himself.

A better strategy for James would be to engage in self-competition. The brain circuits that are activated when we perceive others' status are actually activated when we compete against ourselves. This means that James can receive little shots of reward every time he achieves something new. The achievements can be really small things, such as getting to the meeting first, making a good suggestion, taking on a responsibility that will stretch him, validating a co-worker, or many other things. Ben can help James do this by helping to boost his status through setting him small things to be responsible for during meetings, sincerely telling him he has a good idea or done a good job and checking with him what he has achieved (asking for his successes, so he only has to share the things that boost his status).

Mood and motivation

THE POWER OF STATE

A great experiment looked at the effect of mood on two identical twins. One twin was submitted to a series of situations that created a positive happy mood within her. The other got the raw end of the deal and experienced a negative mood. She was made to go to a quiet room in a gym near a shopping centre. She had to read and sign a serious-looking consent form. She had to read statements from a laptop screen and to imagine saying things to a friend that were designed to make someone feel bad (such as 'I sometimes feel so guilty about the hurt I have caused my parent'). She was also subjected to sad music, Barber's Adagio for Strings. After then being left alone for a while she was taken to the shopping centre and allowed to shop for 30 minutes.

The lucky twin was put through a similar series of situations but all of hers created a positive mood. The statements were, for example, 'I feel I can do just about anything' and the music was uplifting. She was also then sent shopping for 30 minutes.

The results were that the sad twin bought less stuff; she went into fewer shops and looked at fewer items. She also felt less satisfied (not really liking the trainers she had bought). After only 20 minutes of shopping she became distressed and didn't want to do any more. The happy twin bought lots more, went into more shops and looked at more items. She even bought a couple of presents for her sister!

This is a dramatic demonstration of the effect of states on behaviour. Imagine the different results you could get with a client in a positive versus a negative state.

Expecting to be in a positive state all the time isn't realistic or likely to be healthy. Being able to elicit the states you need on demand is very useful. Getting to know what works for you just takes a little reflecting. For example:

- What music makes you feel upbeat?
- What memories can you replay in your mind that get you feeling great?
- What can you say to yourself that makes you laugh or feel invincible? (Or makes you feel any other way you'd like to.)
- What is so important to you that when you think about it you become incredibly determined?

Our thoughts really affect our states. Knowing which thoughts you can draw upon to elicit different states in yourself can give you power and flexibility at the times you need them.

MYBW top tips for motivation

- Identify carefully the times to use extrinsic motivation.
- Strategically encourage thorough intrinsic motivation – ideally recruit people with purposes beyond money for doing their job.

- Ensure that you know what your higher purpose is.
- Give certainty and enhanced confidence whenever you can.
- Practise your own state elicitation so in time you can call on optimal states for any task.

MYBW top benefits for mastering motivation

- You will be in control of increasing your personal productivity.
- You have the potential to increase others' productivity.
- Your frustration for you and your colleagues will decrease.
- You will experience less stress with children and partners by understanding how to motivate them effectively.

10
The resilient brain – did you receive yours?

Kate had endured the day from hell in her mind. Even though it was drawing to a close she still felt all the pent-up frustration and irritation that was lingering from the day. She hoped that talking to Stuart would help her gain some clarity around things and most importantly, understand how to make sure the day didn't repeat itself.

Kate explained that the expectations were relentless. It used to be the case that they would have busy periods and everyone would just put in the extra hours and deal with the stress. Families would know that was intense at work and they'd just have to cope with a frazzled version of their loved ones until the pressure eased off. Then they used to have a few months to recover before the next onslaught. Now though, things had changed. It seemed like the board thought they could just keep up this pace the whole time.

HR keep talking about resilience. Kate sounded exasperated. She keeps hearing that in changing times we need to be resilient. Have grit. Look after ourselves. Yet there never seems to be time to do any of those things. How do you even switch on resilience? Is it something you're either born with or missing?

Some people seem to still have never ending bags of energy and just keep handling everything that is thrown at them. Although, one of Kate's colleagues in her team has just been signed off with stress and anxiety, and she is the least likely person you'd have thought of to suffer with something like that.

Kate feels the pressure. She feels a responsibility to the people in her team and doesn't think she can protect them any more. Many of them are passionate and capable but are also just wrung-out. She had asked for help and support from HR and they organized a half-day workshop on resilience.

Situation

Stuart checked he understood the key issues:

- Kate felt that things were relentless.
- She didn't understand what resilience was or how to build it.
- The drive to protect her team is strong, skills and perceived ability are weak.

This chapter is about what resilience really is and how to build it. It is also clear on what it is not, and what it can't cover up.

Essentially this was going to be a tough session. There was a lot to this situation and it wasn't going to be solved overnight. On the surface there also appeared to be many things outside Kate's control. But Stuart knew there were also lots that could be brought into her control and that would make positive differences.

Stuart started by fleshing out why this was so critical for Kate. Why are organizations so keen that people have resilience? The timely thinking may come from almost every company because they are vocalizing that they are in a time of change. Uncertainty is prevalent. Employees are needed who are open to changes and can deal with the stress, novelty and uncertainty of change. There are certainly links between resilience and adaptability.

The thinking also includes the understanding that higher levels of resilience reduces the likelihood of chronic stress in individuals. It is well documented that chronic stress is bad for physical and mental health. It builds up over time and can reduce productivity and increase sick days.

We are in an interesting time where understanding and cultures are developing. On the one hand, awareness around mental health is increasing. Some of Kate's colleagues have been through mental health First Aid training, and loved it. People acknowledge that mental health is important and that stress can negatively affect it over time. On the other hand, 'crumbling under pressure' is still often linked to an individual being weak and not up to the job.

Stuart shares some scientific underpinnings for resilience which, although subtly, deeply challenge how many of us talk about resilience and coping with stress.

WHAT IS RESILIENCE?

Building resilience is a slow and continual process of brain development which equips people with a series of behaviours and thought processes which, overtime, strengthen their core ability to overcome and prosper from the negative and difficult events faced in daily life.

It is shaped every day.
It isn't fixed.

It is dynamic.

The first thing to be clear about is that there is not one single way of a brain being resilient. There are at least 4 different types of resilience. Within those categories there are many different networks that can be strengthened within the brain that would result in the experience of increased resilience.

Resilience engenders a mental and emotional strength which impacts all aspects of daily work, good and bad. With healthy resilience, you don't just survive in the face of adversity, you thrive off it. It does enable you to bounce back from challenging events more rapidly, returning to full potential, regardless of the difficulty faced.

In theory, anyone can build resilience. They can develop this strong baseline from which stressful events, change and adversity can be tackled. This reduces the physiological and mental consequences of a stress response, which helps with the ability to perform under pressure.

We do know that a resilient workforce that thrives under pressure can deliver more. So, what holds us back? It isn't just the structure of individual's brains. The environment they exist in is also a major factor. What is the culture like? One common source of stress in the workplace is interpersonal conflict. It has a negative impact on teams and can create a hostile work environment. Employees who have greater resilience to situations involving conflict and who are better able to deal with stressful encounters without letting their emotions spiral out of the control, minimizes the emotional fallout and allows the organization to maintain an open and collaborative working environment.

TOP MISTAKES IN THIS AREA

Synaptic Potential has worked with many organizations on this topic and these are some of the classic mistakes we've come across.

1. Dealing with stress and building resilience in the long term are not the same things. They each need a slightly different tailored approach to maximize their benefits.

2. Resilient but not invincible. Just because you have a group of resilient people doesn't mean that they become invincible. They are still human beings, albeit ones who are better able to cope with stress and challenge in an emotionally, cognitively, socially and physically resilient way. But there will still be a limit to this resilience. Expecting them to perform like machines without limits is unrealistic and undesirable.

3. Leaving it to the last minute. Resilience is a process, not a state. This means it takes time and effort to build, and isn't something that you can gain overnight. You have to start early and sustain your efforts so that you can reap the greatest benefits for yourself, and for the organization.

4. Believing that all stress is bad for you. It is well established that too much stress is bad for your health, especially when it is chronic. But too little stress can also be bad for business because individuals lack the drive that they need to push themselves forward. Instead, finding a low levels of stress that can motivate you to deliver your goals in a time efficient manner is the goal.

5. Thinking that stress just affects your body. Feelings of stress are most commonly associated with a host of physiological sensations such as sweaty palms or a racing heart. But stress doesn't just affect your body. It affects your brain too. For example, stress can impair your ability to remember essential information needed at that particular moment. Reducing stress isn't therefore just about alleviating these physiological sensations, or supporting good health, it is about making sure you are thinking at the top of your game.

6. Thinking that one size fits all. There are many ways to build resilience and deal with stress in the moment. What works for one person may not work for another. And what works in one situation may not work in another. You need to take a tailored approach.

7. Just ignoring it. Passive coping strategies, while less effortful are less effective that active ones. Trying to ignore, avoid or suppress your stressful feelings doesn't usually work. Instead, selecting an appropriate active coping strategy.

Explore the short easy ways resilience can be built

One important strategy for building resilience does include reducing stress in the moment. This alone isn't enough – but not doing it at all means you miss a trick. Giving the brain a break from continued pressures and challenges gives it a chance to recuperate. This pushing pause moment is important to keep cortisol levels at reasonable levels. If we imagine our stress chemicals travelling a rollercoaster daily then the ups aren't a problem in themselves, so long as they come down again too. Otherwise the rollercoaster isn't much fun.

Mental rehearsal

For a long time now we have known that mental rehearsal has real world results. Jessie was introduced to this concept in Chapter 8 to help her achieve her goals. This is another of many studies that confirms the exciting reality of the power of mental rehearsal.

MENTAL STRENGTH

Erin M Shackell and Lionel G Standing from Bishop's University tested whether mental training can produce a gain in muscular strength. They got thirty male athletes from different sports together and divided them into three groups. One group did only mental training; one group only physical and the third group did neither. They were aiming to improve their hip flexor muscles and the results showed an increase in 24 per cent for the mental training group, 28 per cent for the physical training group and no significant change in the control group.[2]

Kate is really busy. She doesn't feel she has time for a yoga class in the middle of the day (even though she'd love to). However, she could make time to mentally travel to the place she feels most relaxed in the world, South Africa. She could certainly take 2 minutes to just close her eyes and picture sitting, overlooking a beach with a glass of chilled wine in her hand, feeling the sun on her skin.

Stuart is keen to equip Kate with several other evidence based ways to help her build her resilience.

THE POWER OF NATURE

Music is one way to allow people to experience nature without having to leave their desk (which isn't always possible). These sounds must be captured from the naturally quiet environment rather than silence or anthropomorphic noise.

In a 2013 study, participants exposed to a simulated forest environment with nature sounds had a significant reduction in cardiovascular stress markers and cortisol levels following an induced stressor.[3] This is hugely exciting! These 'booster breaks' are one of many mini-interventions that can be built into the working day and help keep the brain in good condition.

The research has showed that sounds of nature such as wind, water, and animals, are preferred over anthropogenic sounds such as traffic, recreational noise, and industrial noise. A virtual reality forest including sound was found to improve stress recovery more than the same forest without sound. Rural soundscapes and botanical gardens were preferred over urban park soundscapes, which were preferred over urban soundscapes.

Combination of visual nature scenes with auditory stimuli was more effective in managing acute pain than just using one of these methods.

Passively viewing urban parks or woodlands produces greater physiological changes toward relaxation, positive emotions, and faster recovery from attention demanding cognitive performances than watching built environments without natural elements.

Participants in an open-plan office laboratory worked for two hours in noisy setting, consisting of mobile tunes, telephone ringing and telephone conversations. After that, they took a 7-minute break in four different restorative conditions: 1) nature movie with sounds of streaming water, 2) river sounds only, 3) silence, 4) continuation of office noise. The participants who saw the movie with sounds of water rated themselves as having more energy compared to the other three groups.

In a study using mobile electroencephalogram, walking in a quiet natural environment was associated with neurological patterns in line with attention-restoration mechanisms described by Kaplan and Kaplan in the Attention Restoration Theory.

Kate recognizes that her immediate response is 'who has time to go for a walk?'. But she catches herself and realizes that she didn't used to believe this. When her children were young she would always value taking them for walks. She knew instinctively the benefits of being outside and exposure to nature. When did she become dictated to by her work culture? Just because something was the norm there, like working for hours on end cooped up, doesn't make it the best way to work and live.

Stuart was thrilled Kate was starting to question and explore in this way – and it offered a perfect segue into another concept;

CURIOSITY AS A DEFAULT APPRAISAL SYSTEM

This proposition combines two concepts:

1. How we appraise situations is critical to how we respond to them

2. Curiosity is a powerful and safe lens to appraise through

Curiosity is brilliant because it does two things:

- It gives you the opportunity to learn and discover something new.
- It causes you to overcome your in-built fear of the unknown.

According to researchers at the University of North Carolina, feeling 'interested' is the emotion that is associated with being curios. When you are curious about something you evoke a feeling of 'being interested'.

Curiosity motivates you to try something new – maybe a new dish in a restaurant, or a different genre of film to what you normally choose. But in doing so, it causes you to challenge your brain to shift its perspective from preferring to stick with what is familiar, towards going out and experiencing something that is new.

Being curious satisfies your desire to discover interesting and new things to explore. Without curiosity your need to be interested is not properly satisfied. And although feeling 'interested' isn't what you would consider to be one of your core emotional states, it is a feeling that we all experience and is what is classed as a 'cognitive' emotion – one that requires a more complex pattern of thinking and emotion – something that human beings are typically very good at because of our highly developed prefrontal cortex.

Without the feelings of 'interest' and the state of curiosity, change, novelty and uncertainty can all evoke fear. The brain generally prefers to stick in the safe territory of what it already knows – the so called familiarity heuristic – or familiarity bias. And that is one of the contradictions of your brain. While on one hand it loves to learn something new, on the other hand it prefers to stick with the familiar, and that 'something new' makes it slightly fearful. This makes sense from a protectionist evolutionary perspective, but can also carry over and act as a barrier in seemingly non-threatening situations simply because of the presence of something that is unknown or a bit different.

To be curious you therefore have to overcome this fear of the unknown, the fear of something new, something that is challenging to your status quo and decide to give it a try.

Interest and curiosity are personal. As stated by Paul Silva from the University of North Carolina, 'one person's dissertation is another person's indifferent shrug'. But what you are curious and interested in is not fixed. It is context dependent – the people, the place, the time in your life and your experiences can all influence what you are interested in and therefore what you are driven to be curious about. That means that what you might find scary at one point in time, you might find interesting at a later point in time. It also means that any task, however challenging it may seem, has the potential to be interesting to you.

Kate immediately gets that this idea of curiosity as a default appraisal system is profound. While it sounds straightforward, it could change everything. She had recognized from Stuart during previous Coaching sessions that the power of how we label things shapes our experience of things. When she talks about a meeting she is dreading that affects her brain chemistry and primes her in a particular way. The micro-appraisals she makes daily are equally powerful. Stuart agrees that we can imagine all these micro-appraisals chiselling away at our brains, shaping them. The question is, are they #buildingbetterbrains?

The research around mindfulness is very persuasive and more commonly accepted. However finer nuances of the opportunities available to us are frequently missed.

MINDFULNESS PLUS

There are various types of mindfulness (eg focused, empathic) and different types have different effects on your stress response. In addition mindfulness programmes can be variable in length and vary from three days to nine months. This is why some of the literature on the effects of mindfulness on stress is inconsistent, especially in terms of whether mindfulness can reduce the physiological response (eg cortisol) to stress. Selecting the right kind of mindfulness training is therefore dependent on what works for you, and also what suits your lifestyle best as it is best performed as an on-going habit, rather than a crash course. In addition, some forms of mindfulness training are effective at decreasing feelings of stress, but are less good at being able to elicit physiological shifts in the stress responses (important when thinking about stress in the moment). For example, 3 day or 8 week programmes have been shown to benefit self-reported stress levels, but are more variable in their ability to reduce cortisol responses in the moment and improve wellbeing.

In 2017 Veronika Engert and colleagues did a very interesting study looking at reducing cortisol levels. This large-scale study showed that 3 months of attentional-awareness mindfulness was effective at reducing self-reported feelings of stress, but did not have a complementary change in the stress physiological response. In contrast, performing 3 months of mindfulness training techniques which focus on emotions, especially compassion and empathy or perspective taking, reduced both self-reported stress and cortisol-based stress reactivity by a third. In addition, when performing the attentional-awareness mindfulness training programme prior to the emotional/perspective taking mindfulness training programme, it further boosted the benefits on stress reduction by nearly 50 per cent. This shows that combining multiple types of mindfulness is important for helping to maximize the reduction in stress response.

But it is also important to remember that any level of mindfulness can help to some degree, as people low in mindfulness – in other words people who don't practice mindfulness – are typically those who show the greatest psychological reaction to stress in the moment. Brain imaging studies have also shown how this type of training has an effect on the structure of your brain, by increasingly plasticity and flexibility in brain regions involved in social and emotional thinking – an example of how your behaviour can actually change your brain as you develop and improve.

Other evidence for the benefits of mindfulness suggests that it can be particularly effective if combined with exercise, showing that it doesn't have to be done sitting down but can also be integrated into your daily routine.

Can you increase your mental endurance?

Kate seems more positive. She has seen possibilities and how small changes could really benefit her and her colleagues. Genuinely caring for those in her team means she really does want things to improve but she hadn't been sure how. There is a remaining fear though. What if the senior leaders just keep expecting more and more? What do we do if the bar just continues to rise?

Stuart explains some of the science behind why continuous bar raising isn't a good plan.

COGNITIVE FATIGUE

Cognitive fatigue is defined as a decline in the efficiency of your mental abilities that is caused by excessive mental activity.
It can have several causes. One of these is simply the act of having to sustain your attention over a prolonged period of time on a particular task or tasks.

Another is a disrupted night sleep which means your brain hasn't had sufficient night-time recovery time to ready itself for the new day. High levels of stress and anxiety can also cause exhaustion and mental fatigue.

Fatigue is associated with a sense of discomfort, a desire to rest, and a decline in motivation and task performance. These sensations work as a biological alarm which urge you to take a rest so that your neural homeostasis can return to normal and you can recover from the fatigue.

Cognitive fatigue most commonly leads to a failure in your attention systems. This ultimately means that you can't focus properly on the tasks that you are meant to be working on. You are less able to attend and focus properly on information coming in from your surroundings. You are less able to multitask effectively because you find it difficult to divide your attention between two things at once. And you are less able to inhibit unwanted, interfering and irrelevant information causing it to overwhelm and clog up your thinking.

COGNITIVE PERFORMANCE

The precise mechanisms behind how mental fatigue causes a detriment to your cognitive performance are still not fully understood but there are a few pointers which are starting to show the way. One suggestion is that your executive control system – the system which decides where to allocate your mental resources – is impeded or compromised when you are mentally fatigued. This leads to more disorganized thinking where you are less able to direct the mental resources where they are needed to complete the task and less efficient at switching your internal mental focus from one cognitive process to another. Alternatively, others have suggested that it could be that there are metabolic changes inside the brain caused by the sustained mental activity (similar to how there is a build-up of chemicals in your brain during the day which provide a signal for your need for sleep) which both signal your need for rest and disrupt the neural functioning the brain, making it work less efficiently.

But why is it that you can sometimes battle through mental fatigue? Well some scientists have recently proposed that your brain has two systems which help to regulate the effects of cognitive fatigue – a 'facilitation' system and an 'inhibition' system. Your facilitation system helps to compensate for mental fatigue so that you don't get the same performance decrements that you might otherwise suffer from. In contrast, your inhibition system acts to mediate the performance decrements of mental fatigue. The balance between these two systems therefore determines how much your performance will be adversely affected by your fatigue. But these systems can only operate over the short term, and long-term activation of your facilitation system leads to chronic (as opposed to acute) fatigue which can have detrimental effects on your mental and physical health.

In other words, at some point you just need to rest and there is nothing more the brain can do to help maintain an efficient state of thinking.

Organizations say that they want to take care of their people. Their wellbeing is important. Some even say their fulfilment is important. Often really great companies fall down when they say one thing and act in a different way. A company that exists purely to deliver ever raising profits can find itself conflicted. Messages are communicated through formal internal communication channels – but speak higher volumes when demonstrated in leader's behaviours.

BURNOUT: A DESTINATION OR A JOURNEY?

Burnout is an extreme and chronic form of mental fatigue, with the added element of you suffering from excessive levels of stress. It not only causes exhaustion and an impairment in mental functioning, but it also results in behavioural changes such as increased irritability, frustration and cynicism. It ultimately may also cause you to increasingly distance yourself emotionally from others, and start to feel numb about your work. The fact that symptoms of burnout can be so severe, long-lasting and share similarities with conditions such as depression, has led scientists to debate whether it should be considered a mental health disorder rather than a symptom of mental overload.

But while burnout is the rarer extreme of mental fatigue, everyday mental fatigue is usually short lived and manageable simply by making sure you give your brain the rest and down-time that it needs.

Another consideration is around emotional intelligence. A lack of emotional expression is often mistaken for resilience. Resilient people often have emotional intelligence – but this definitely does equate to being emotionally muted.

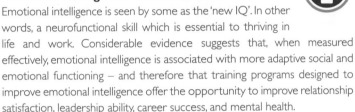

WHAT IS EMOTIONAL INTELLIGENCE FROM THE BRAIN'S PERSPECTIVE?

Emotional intelligence – back to basics

Emotional intelligence is seen by some as the 'new IQ'. In other words, a neurofunctional skill which is essential to thriving in life and work. Considerable evidence suggests that, when measured effectively, emotional intelligence is associated with more adaptive social and emotional functioning – and therefore that training programs designed to improve emotional intelligence offer the opportunity to improve relationship satisfaction, leadership ability, career success, and mental health.

However, the extensive discussion around emotional intelligence has resulted in a huge number of different models, theories and ideas which form a disorganized landscape of knowledge that at times is hard to disentangle. It is a personality trait? A cognitive ability? Or a form of emotional regulation? Or a combination of all of these? And so if we pull back emotional intelligence to the level of the brain, what does it actually look like?

Defining emotional intelligence

If we go back to basics then emotional intelligence, from a literal perspective, can be seen simply as the combination of emotion and intelligence. This idea was built into early definitions of emotional intelligence in the 1990s which stated that emotional intelligence is the ability to 'monitor one's own and others' feelings and emotions, to discriminate among them and to use this information to guide one's thinking and actions'.[4] If you break down this definition, then you can see it splits into several main themes which include the ability to recognize and understand emotions in yourself and others, the ability to effectively regulate the emotions of yourself and others, and the ability to adaptively generate emotions in yourself. All of these concepts have been covered to some degree in neuroscientific study and can therefore help to inform our understanding of emotional intelligence.

Tracking emotional intelligence

One challenge to the investigation and application of emotional intelligence lies in the difficulty of knowing which are the best way to measure it in the moment, as well as being able to track it over time to monitor potential improvements. Debates as to whether self reports or performance measures should be used are frequent and still unresolved, potentially explaining the vast number of metrics available to individuals and organizations looking to measure emotional intelligence, as well as the variability in results on offer across the scientific literature. Taking a neuroscientific perspective allows us to bypass some of this debate and get to the crux of what emotional intelligence means at the level of the brain.

Emotional attention

Although we often think of attention as our ability to concentrate, the selective nature of attention is another aspect of it which can be both an advantage and a disadvantage. For example the ability to efficiently recognize emotional states both in ourselves and in others requires that we actually notice them – ideally at a conscious level, rather than ignore them. This process of emotional attention occurs not only towards the sensory cues (eg facial, bodily, vocal) that we pick up from the people around us, but also in terms of noticing our own internal bodily state which provides useful signals for emotional recognition. It is often assumed that these basic skills are obtained instinctively through day to day experience, but this may not always be the case. Providing this kind of training alongside opportunities which ensure that the individual has the right conceptual and linguistic knowledge about the wealth of emotional states, means that they can be better equipped to recognize emotions both in themselves and in others.

Empathy, perspective taking and reappraisal

Emotional recognition is only one part of emotional intelligence. Regulating and expressing emotions are also critical elements. While regulating emotions in others is often covered by interventions which encompass ideas relating to empathy, regulating emotions in oneself often requires the individual to practice good emotional habits such as using effective reappraisal strategies, harnessing the benefits of effective perspective taking and inhibiting maladaptive behaviours which they might do without even thinking about it. All this can help someone to manage their inner emotional flux which at times can overwhelm, and make sure that they offer an appropriate – and intelligent – expression of emotion, especially when under challenge.

In summary, a brain's perspective on emotional intelligence allows us to get to the roots of how emotional intelligence maps onto our brain's inner workings. This approach allows us to utilize neuroscience-based insights to develop effective strategies and solutions which are concretely aligned with the way the brain works, rather than simply operating within a theoretical landscape.

It is right to be aware of neurodiversity. Not everyone will be highly emotionally intelligent, and nor do they need to be. However, people working together in teams and people responsible for managing or leading others may benefit from particular skills. More attention needs to be given if certain skills are missing. A productive and happy team is possible – it just will likely require more thought and trust and communication.

Challenging perspectives

Kate thinks their culture is good but they're definitely not in the place where people can openly discuss their emotional skillset and share what they are personally working on. She brings up her colleague who recently went off sick with anxiety again and Stuart probed her thinking around this.

Essentially, Kate explains, we thought this lady was tougher than she was. We thought she would bounce back from the challenges and the pressure. It's now come to light that she was having some challenges with her teenage daughter at home, and I know her mum has dementia. But she always seemed so capable. Perhaps she just wasn't as resilient as we all thought Kate concluded.

THE RESILIENT SNACK COMPANY

Synaptic Potential were asked by a global snack company to help them strengthen their 'already good' resilience training for staff. They stood out as a company because they weren't fixed on solving resilience within a two-hour workshop. (You'd be surprised how many requests are made for magic!). They were serious about supporting their people and saw value in sharing the science behind how they could build resilience over time.

This company have an amazing online university, so employees from around the world can access well designed modules about a huge range of topics. To this they added the series of neuroscience underpinned exercises and tools individuals and teams can use to build resilience and cope during times of stress. The whole programme went far deeper than a single touch point, and this is necessary for long term brain changes.

More companies need to step back from doing a one day training on a requested topic and supply well researched approaches and content that really help people.

MYBW Top tips for resilience:

- Develop daily practices for building resilience.
- Spend time in nature.
- Practise curiosity.
- Explore mindfulness.
- Take daily stress management steps seriously.

MYBW Top benefits for mastering resilience:

- Good employee wellbeing.
- Increased productivity.
- Enhanced engagement.
- A host of benefits for outside of work (marriage, friendships, parenting) which influence performance in work.

11
What to do when everyone demands innovation

J essie had a team meeting coming up. The development and future of the company relied upon people being more creative and innovative. They had to stay ahead of the curve and come up with new ways to solve old problems. While Jessie understood the technical challenges of the problems she was trying to solve, and what solutions they needed, she didn't feel that she was 'creative'. She excelled at science! Surely, she wasn't expected to be both 'sciencey' and creative?

It was the same with her team. The graphics person thought she was creative...but didn't feel she knew enough about their client's challenges to come up with innovative ideas around how to tackle them. The whole team would often get together, knowing that they should be coming up with new ideas, but they'd sit in their office room and try to brainstorm. Typically, they might come up with something mediocre.

Situation

Jessie believes:

- She needs to be creative and innovative.
- This is vital for business.
- No-one in her team feels equipped to innovate.

Stuart explained that there are different stages to creativity. The current thinking is that everyone has the capability to be innovative...we just should facilitate it more effectively. Understanding the different networks involved and how the brain works helps.

This chapter is about becoming more creative. We know now that anyone can be, and the challenge is understanding that and then designing opportunities for your brain to work in the necessary ways to release your creative potential. Rex Jung, a professor of neurosurgery at the University of New Mexico says 'Everyone is creative; it's just a matter of degree. We have this prototypical idea of artistic creativity, but we are creative in our relationships, our work, our cooking or even arranging our homes in a different way.'[5]

Jessie isn't easily convinced. She talks about famous creative people from history: Thomas Edison, Walt Disney, Einstein, Mozart, Leonardo da Vinci. 'You can't tell me I'm the same as them.' Stuart agrees it is unlikely that tomorrow she will create a musical or visual masterpiece. But that isn't the whole story. For sure most of us have not exercised some of the skills required for creativity... but they can be resurrected.

WHAT IS CREATIVITY AND INNOVATION?

What is creativity? It could be defined as productivity marked by originality.

It involves mental processes such as decision making, language and memory. We may need to set aside our normal ways of viewing the world or allow unconscious thoughts to bubble up to the surface of consciousness.

Stage 1: There are different stages to creativity. Idea generation is the most well-known, and is typically the first stage. During this stage we benefit from having lower cognitive control. Few restrictions on what you come up with or how. If someone wants to play on a swing or lie on the floor while generating ideas; all good. Being in a state of hypofrontality (lower activity in the prefrontal cortex) offers creative benefits. This state of lower cognitive control is associated with alpha waves around 8-12 hertz and are a sign of relaxed wakefulness and diffuse attention.

Stage 2: This stage involves evaluating the options that have been generated. This typically culminates in selecting an idea to brainstorm implementation considerations for. Here we need the cognitive filter in the prefrontal cortex switched firmly back online. This enables people to critically appraise ideas and put the brakes on.

Reflect on the times where some people in a group are in stage 1, while others are clearly in stage 2 and vocalizing their evaluations!

Myth busting

Creativity is 'located' in one hemisphere or area of the brain.

Reality: creativity is a whole-brain activity.

Some people can only do stage 2.

Reality: they may think that, and may only choose to exercise those networks in their brain, but stage 1 capability is there... it may need some practice though!

Core components of creativity

Rex Jung encourages us: 'The more raw material you have, the more time you devote to developing a skill set, the easier it is to improvise. It takes expertise to have enough material to draw on to be creative. So find an area that interests you, develop an expertise in that area, and then start creating and develop something extraordinary.'

1 Capture your genius (or tragic ideas).

2 Expand your exposure.

3 Challenge yourself?

4 Be intentional with your surroundings.

Improvization

Stuart was really unsure how Jessie was going to take today's session. He knew she was onboard with re-wiring her brain, and that at times that meant inviting her out of her comfort zone... but this may be a step too far for her.

He starts by explaining the benefits and science behind why he is proposing this kind of activity. Innovating new ideas, especially with other people around, can involve being vulnerable. Putting yourself out there to be examined and critiqued by others. This can be tough. It can prevent gems of ideas from being tabled and improved upon or acting as stimuli for others to pivot off.

People whose brains have adapted to this are improvisers. Typical individuals to study include jazz musicians, ideally piano players (since you can get a keyboard into an MRI scanner...but not so easily a tenor saxophone). When someone is improvising (which is essentially making up music on the spot, either individually or in response to another musician) specific things happen in the brain. Charles Limb, a neuroscientist and also accomplished jazz musician himself, has done great research in this area.

Part of the prefrontal cortex, called the lateral prefrontal region, shuts down. These areas are responsible for our self-inhibition, our self-consciousness, our ability to judge how right or wrong what we're about to do is and control this. In contrast, the medial prefrontal cortex (the default network) becomes more active. This network is key for self-expression and our autobiographical narrative. So essentially our brain is shifting into a zone where we can let go of any external judgement and are free to express ourselves which is fantastic for the first stage of creativity!

Other research has shown that the other benefits to practising musical improvisation include being able to express yourself and communicate with others, facilitating self-actualization and deeper connection with others.

Jessie is invited to give it a go. There is a selection of instruments for her to try and a variety of types of background music to give some context. Home play is to attend a jazz club and have another go at home.

Cognitive flexibility

Cognitive flexibility is a winning skill for so many things in life. While the typical technical definition is broadly around the ability to adapt behaviours in response to changes in the environment – this is more from a behavioural angle. This is critical, and we can add a purer neuro angle which, for creativity, covers being able to up or downregulate the cognitive control system.

Being able to let go of existing views to change your perspective based on updated requirements is really useful. It paves the way for creative thinking. Similar processes involved in counterfactual reasoning enables people to use their imagination to consider past results and possible alternative outcomes to plan the future.

STROOP TASK FOR CREATIVITY

In 2010 Daray Zabelina and Michael Robinson, at the time at North Dakota State University, looked at a group of students. First they assessed their creativity through standard paper and pencil tests. Next they got them to undertake the Stroop task. This task measures how well you can filter out irrelevant information and focus on important details – a big feature of cognitive control.

Interestingly overall, those who were currently assessed as creative and those who were less so all performed similarly. However, 'creative' people did better each time they had to switch from a matched option (eg the word *red* shown in red) to a non-matched option (eg *red* being shown in the colour blue). This stronger cognitive flexibility supports the ability to generate novel ideas, and put them into action.[6]

Mental workouts

Jessie is clear that she wants to train her brain more to strengthen herself in ways of thinking that she has been weaker on after years of studying and working the way she has been. She pushes Stuart for more exercises she can do at home.

Alternative uses – take an object, like a bowl, and generate as many different uses for it as possible within 3 minutes.

Many alternative uses – take 12 different objects and generate up to 6 different uses for each of them in 15 minutes.

Unusual descriptions – try describing things generically. For example a candle described as wax and wick, or even better, string and cylindrically shaped lipids.[7] Challenge yourself; can I break the description down more? Is my description implying a specific use?

Different order – do something you frequently do, in a different way. For example, make a sandwich in a different way, take a different route to work, do your workout in a different order.

CREATIVE PROBLEM SOLVING

In this experiment some people were asked to do the 'many alternative uses' challenge. Next they were asked to solve practical problems, such as trying to fix a candle upright on a wall using a book of matches and a box of tacks. Those who did the many alternative uses challenge first solved many more of the practical problems than those who didn't do the challenge first. They were primed for problem solving.

In 2012, Tony McCaffrey from the University of Massachusetts Amherst, trained students to use 'unusual descriptions' exercise and then were evaluated their problem solving. Those who had received the training demonstrated a 67 per cent boost! The suggestion is that this is because they were more likely to notice obscure features of the problems that then helped them solve it.[8]

Also in 2012, Simone Ritter and colleagues of Radbound University Nijmegen invited students to prepare a classic Netherlands breakfast sandwich of butter and chocolate. Half did this in the normal order and half in an unusual sequence of steps. Next everyone did two activities. Generate as many uses for a brick in 2 minutes and come up with as many ideas to answer the question 'What makes sound?' in 2 minutes. The unusual sandwich group generated more different ideas and scored higher on cognitive flexibility.[9]

Jessie is always looking for the little tips she can use immediately, especially with her team. She loves it when she does something Stuart has introduced her to and the team gets better results that very day! He has another of those very tips lined up. Sometimes the ways we work seem a bit crazy and hard to believe. But if they work…

NEAR OR FAR

Lile Jia and a team at Indiana University looked at how people solved practical problems. They experimented with the framing of this challenge. Some people were told that their responses would be shared with scientists at a university a few thousand miles away while some others were told the results were for a team at their own university. The third group of people weren't given any frame around who the data from the study was for. Fascinatingly, it was the people who thought they were solving problems for the researchers at the far away university who solved twice as many problems as the other students![10]

The thinking is that the psychological distance enables people to approach things more abstractly and this helps find solutions.

Stuart talked to Jessie about how this could work in her context and she came up with several ways she could apply this strategy in their team problem solving. They explored how this concept works for time too. Distancing yourself, imagining yourself one year instead of one day in the future, can also help to solve more problems.

Her team seemed to be falling into the classic brainstorming trap of getting together before individual idea generation. Stuart shared that the latest research shows that actually a relaxed semi-structured scenario, like a lunch, is better than long, formal meetings for when people do come together to brainstorm.

INNOVATIVE TEA

Twining's are world renowned tea makers. Their history is beautiful and their dedication to a great product is impressive. So I was personally excited to see their new offering of a cold infusion. This bag is designed for cold water, especially to go in water bottles. Since I don't drink tea, this seemed a great addition to their range. But I was curious...having worked with this great brand on other things... had they learnt anything during this big innovation leap? An innovation that many others quickly copied the end product of.

Some of their reflections included creating a dedicated team to work on projects like this to fast track them. In this instance the project was one of many which slowed things down. Ideally they'd also recommend having people on the project that can see it from the start to the end. Losing key people during an innovative process can be undesirable.

The strengths were that they started with their customer's needs. They knew they are great at adding things to water to liven it up...but why were they wedded to hot water? They aspired to 'make water wonderful'. So they played with ideas. Their ambition was important to them and creating this new space, new category with a desire to be 'game changing' meant they remained honest and bold.

Brain during creativity

Divergent thinking, commonly measured by the alternate uses task, is all about how laterally people can think. Is their thinking locked in, following predictable pathways or able to hop about and make different connections? People who performed better on these tasks seem to report having more creative hobbies too. Jessie shared that currently she didn't make time for any hobbies, but in the past she had enjoyed Chinese calligraphy.

ALTERNATE USES TASK

This common experiment involves asking people to come up with different uses for items. It was mentioned briefly in 'Mental Workouts'. For example, uses for a sock. A common use answer would be to warm your feet. An uncommon answer might be to use the sock as a water filtration system.

Roger E Beaty and colleagues did some research around a 'high-creative' network to see if someone with stronger neural connections in this network would score well on creativity tasks.[11] Overall they did, those with stronger connections came up with better ideas. So what are these brain regions involved in the 'high-creative' network? The brain regions belonged to three particular systems:

Default network – activated when people are not focused on a task, for example daydreaming, imagining, mindwandering. Important for idea generation or problem solving.

Executive control network – important for focus and controlled thought processes. Important for evaluating ideas.

Salience network – regions that switch us between default and executive networks. Important for transitioning between other networks and also bringing things into our attention.

These networks are not typically active together. It may be that very creative people have become good at co-activating brain networks. We know from various researchers work (de Manzano et al, 2012b; Limb and Braun, 2008; Liu et al, 2012) that the dorsolateral prefrontal cortex has been shown to activate and not activate during creative processes. Ana Luisa Pinho and team suggested in 2016 a mechanism for cognitive creative strategies depending on the context. It appears the brain may use extrospective or introspective neural circuits.

'Wow' exclaimed Jessie. That was a lot of information. She started to wonder how she could co-activate her brain networks.

COGNITIVE DISINHIBITION

In 2003 Jordan Peterson and Shelley Carson[12] found that highly creative individuals are more likely to display cognitive disinhibition compared to less creative individuals. Cognitive disinhibition is the failure to ignore information that is irrelevant to survival or things you're wanting to focus on in the moment. Neuroscientists typically speak on cognitive inhibition as a good thing to prevent distractions. If we didn't cognitively inhibit information then with the vastness of data bombarding our senses we would be overwhelmed.

The research suggests people who have higher scores on divergent thinking tasks, openness to experience trait, the Creative Personality Scale and the Creative Achievement Questionnaire also tend to have lower scores on the latent inhibition task. Peterson and Carson thought that this lower cognitive disinhibition means more data can get into conscious awareness to be reprocessed resulting in more creative ideas.

At school Jessie was always very good at focusing. She was considered a good student. Did that mean she would be good at cognitive inhibition? Stuart said that it might do. But, with neuroplasticity you can always practise not inhibiting. Jessie could let her mind wander more often and see what happens.

Cognitive disinhibition is also closely involved in the 'aha' moments. These are often what people associate with creative insight, but in reality are only one of many routes to innovation. When we get these moments of insight it means our cognitive filters have relaxed enough to allow ideas deeper in the brain to be brought into conscious awareness.

'AHA!' MOMENTS

John Kounios and Mark Beeman are well known for their 'Aha!' Moment research. In their experiments they asked people to solve word-association problems while their brain patterns were recorded using fMRI or EEG. The individuals would indicate when they got the answer and also whether it had been through a flash of insight or more of a trial and error process. From this the researchers could see that a period of alpha activity precedes a burst of gamma activity at the moment of insight. Their conclusion was that alpha activity focuses attention inwards and the gamma burst occurs as the solution moves into conscious awareness.

Stuart discusses 'Aha!' moments with Jessie. The science is telling us a few really important things. Forming these novel insights often requires us to break down mental representations to bring in new information and make new meaningful connections.

Several different research teams have found that highly creative people seem to produce more brain waves in the alpha range during creative tasks when compared to less creative people. This used to be linked with a defocused attention. Now the thinking is that the brain is focusing on internally generated stimuli rather than the data coming from the outside environment.

Jessie became excited and thought there must be a simple way to increase alpha waves. Stuart shared what he knew and encouraged Jessie to do her own research. There were some studies that showed increases in creativity based on neurofeedback and Heart Rate Variability feedback and at least one study suggested narrower benefits. There was still more to be done in the field though with double-blind studies. Activities that relax you may help to increase alpha waves. Meditation, deep breathing, relaxing baths, yoga or just closing your eyes may help.

Glorious sleep

The value of sleep is huge. It is the gift that keeps on giving. There are ways you can prime yourself to make insights during sleep more likely. This is an inspirational story from the field of mathematics.

SWEET DREAMS

In the 1950s a man called Don Newman was a lecturer at the Massachusetts Institute of Technology. A future Nobel laureate called John Nash was also there at this time. Newman tells of a particular maths problem he had been trying to solve. He recalled struggling and wrestling with it but not making any progress.

One night he dreamt that while he was reflecting on the problem Nash appeared. Newman explained the details of the problem to Nash and asked his advice. Nash explained to him how to solve the conundrum! (All in the dream). Newman woke up, with the answer and invested several weeks in writing it up into a formal paper for publication in a journal.

It is reported that many great ideas have come to people during dreams. Mahatma Gandhi's call for a nonviolent protest of British rule of India. Friedrich August Kekulé came up with the structure of benzene after dreaming of a snake made of atoms taking its tail in its mouth (the structure of benzene is circular).

Stuart explains to Jessie that dreams can be really helpful for problems that require creativity or visualization to solve. Researchers believe that dreams are thought in a different biochemical state. When we sleep the physiological demands on our brain functions are different. Our perception of our thoughts when we are asleep is different, yet we are still focused on issues that are pertinent when we are awake. This can be a huge bonus for us. Our brains are able to explore ideas outside our normal patterns of though. We can even encourage our brain to ruminate, while asleep, on particular issues.

SLEEPING YOUR WAY TO INSIGHT

In 2009 psychologists at the University of California, San Diego, looked at whether REM sleep affected problem solving. People were given a test that required creative problem solving and then given some hints about the answers. Some of the people then remained awake, some in non-REM sleep, and some in REM sleep. Then they all took the test again. Those who got the REM sleep showed the most improvement on their creative solutions!

In 2009 Harvard University's Robert Stickgold's lab was also investigating the effect of REM sleep. Students were given a challenge, which involved weather prediction with a hidden general rule. They learnt by trial and error. Those who were given a nap part way through the process were more likely to discover the rule. In fact how much REM sleep they got directly correlated with heightened performance and ability to explicitly articulate the general rule.

Through other research it has been clarified that only the REM sleep sharpens this type of performance.

Stuart suggests that Jessie recalls particular challenges before she goes to sleep. Just posing them as positive questions before she drifts off. In the morning just spend a few minutes seeing whether you can recall anything from your dreams.

Daydreaming

It sometimes amuses Stuart that in schools daydreaming is typically regarded as a terrible thing, when actually it can be very fruitful at all ages. In our daydreams we can rehearse future scenarios without risk. We can learn from them. When we allow the mind to roam freely it can stimulate creativity. We know that the Default network becomes active when we allow our mind to wander from the current task into travelling to the past and future. As the story goes, daydreaming worked for Einstein who pictured himself travelling down a light wave, helping him create his theory of special relativity.

Some researchers would encourage more daydreaming time. The key for increasing creativity is that we need to pay attention to our daydreams. Your brain needs to be free to wander, but also disciplined enough to notice when you have an idea. Jessie has almost full flexibility in this. She could also allow her team to nap during the day easily. But for those for whom this isn't a reality yet there is something else they can try. When taking a break from a creative problem solving session, do something mildly demanding, like reading. Those that try this instead of doing nothing at all, or a highly demanding task, appear to perform better.

Play

TIME TO PLAY?

We know that for children giving them time and space to play is really important. It positively impacts social, emotional and cognitive development. Allowing imaginative and rambunctious 'free play' in contrast to games or structured activities is critical. Without this the likelihood of anxiety and social maladaptation increases.

What about adults though? Research is showing us that it is important for them too! Marc Bekoff, an evolutionary biologist at the University of Colorado at Boulder suggests we need play to avoid getting burned out from the 'hustle-bustle busyness that we all get involved in'.[13]

Stuart Brown, a psychiatrist and founder of the National Institute for Play in Carmel Valley, Calif, suggests three ways to get more play into your life[14]:

1. Body play – some sort of active movement that doesn't have a time pressure or particular outcome

2. Object play – using hands to create something, again not necessarily with a specific outcome

3. Social play – purposeless social activities

The inspiring Marian Diamond was involved in lots of research during her lifetime that helped us understand the benefits of enriched environments[15].

Jessie started to realize that she hasn't been having playful fun in ages. She used to make time to go and see her niece regularly but recently she has been focusing on work. Maybe she would be better at work if she spent a little more time doing other things. Stuart pressed her to see what else she might enjoy. She used to love ice skating...but wasn't wanting to train for exams or anything now. Perhaps she didn't have to though. Maybe she could just go once a month and enjoy it.

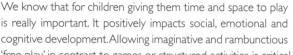

LEGO PLAY

Many companies have embraced Lego serious play. On the front page of the serious play website you can read 'The Lego Serious Play methodology is an innovative process designed to enhance innovation and business performance. Based on research which shows that this kind of hands-on, minds-on learning produces a deeper, more meaningful understanding of the world and its possibilities, the Lego Serious Play methodology deepens the reflection process and supports an effective dialogue – for everyone in the organization.

The Lego Serious Play methodology is an innovative, experimental process designed to enhance innovation and business performance.'[16]

Sean McCusker, an associate professor in education at Northumbria University in Newcastle says:'I have been using the Lego Serious Play method for about five years now – in the UK, China, Malaysia and the US. I've watched people play with Lego in a variety of environments: small businesses, trainee teachers and those working in international research. In every case I have seen the transformation – from dubious playfulness to deep engagement – with participants building models that represent complex, abstract ideas.

Participants tell me how the method has allowed them to express things they would not normally express. How they now have a much clearer understanding of how others see the world and how these ideas relate to each other. And in this way, Lego can help to overcome some of the drawbacks of conventional meetings and discussions.'[17]

Stuart explained that even the novelty and fun of the experience was likely to release dopamine which can help with attention, mental shifting, creativity, motivation and enhancing memory. Playing with Lego, or another medium would encourage a submersive experience which is more likely to engage brain networks for executive control which would make turning down distractions easier. Jessie considered how she could bring something like this into her small team.

'It's not how smart you are. It's how well connected you are.'

Mark Thomas – University College London.

MYBW Top tips for creativity

- If you are struggling, take a break and do something dramatically different from what your job entails.

- Disrupt yourself – mix up your routine, try to improve other people's ideas, set aside a good idea to look for a better one.

- Nap or enjoy some mindwandering.
- Have some alone time to get the creative process started.
- Do different things – try new experiences, hobbies, foods, travel.
- Talk to a wide range of people about your work or challenges.
- Prime your REM sleep, and indulge mid-way through problem solving sessions.
- Have fun, enable your positive state to help link up remote associations in your brain.
- Know your area of expertise well – but explore others too.
- Look at things really carefully, and from different perspectives.

MYBW Top benefits for mastering creativity

- Your contribution to any project dramatically increases.

12

Appearing competent to others

Understanding how you remember things and others forget things

Doing a good job is really important to Ben. He likes to know he is giving his best and ideally he likes people to notice that he is conscientious. While he doesn't like the idea of being proud of himself especially or telling people about his achievements, he recognizes that sometimes in the corporate environment he is in it is the easiest way of people knowing that he is contributing. One day he'd like to be a partner in his firm and for that to happen senior managers need to notice him.

This week there were several things he wanted to improve on. On Monday he had been to a meeting with several of the partners at his firm. He knew it was an important meeting where the project he was working on would be discussed. Before going in he had lots of ideas about how the firm could serve the client even more fully and was looking forward to wowing the partners. During the meeting it just didn't seem to pan out that way though. When questions were asked he felt as if he went blank. He was worried that his memory wasn't working properly.

When detailed questions about the project came up he couldn't remember all the figures off the top of his head. He is worried this makes him look as if he isn't interested in the project or that he doesn't take things seriously. When he realizes he can't remember he tries to see if he can find it in his notes, but ends up scrabbling around producing nothing.

After the meeting he catches up with one of the women in his team and asks her about a paper she had said she would send him. She replies that he had said he would write something first and then she would send the paper, but he didn't remember this at all. This happens to him sometimes, people say that they've agreed things but he remembers things very differently. It certainly happens with Rebecca, his wife, and this is often the cause of some contention.

Finally, Ben also wanted to get some clarity about working with rapidly changing agendas. The other week he e-mailed one of the partners about the firm getting involved with a local charity. He doesn't get any response, but after seeing the partner again at this meeting he decides to send a follow-up e-mail (he didn't manage to catch him in person). The response he gets is curt and almost rude. Ben feels frustrated and as if he cannot win. At other times the partners were saying they wanted people to get more involved with the local community and to be a socially responsible company.

Situation

Stuart planned to address the several things Ben wanted to improve upon. They are:

- remembering things in meetings, rather than feeling blank;
- being able to lay hands on figures quickly;
- understanding how people remember things differently;
- working with seemingly fickle partners.

This chapter is about appearing competent to colleagues by using your mental capacity efficiently. The bonus is enabling you to support teams to maximize their results.

Insights

Ben expects a lot from himself. Occasionally this means he puts a lot of pressure on himself to perform. A high level of pressure can affect performance in various ways. Stuart wants to address Ben's concern about his memory not working properly, and also use it as an opportunity to explore some things that may help Ben in various other situations.

When we want our brain to come up with great ideas or insightful angles on things we are often looking for our brain to have an 'insight'. Mark Jung-Beeman is a cognitive neuroscientist from Northwestern University. He has spent over 15 years trying to work out what happens when the brain experiences an insight. He describes it as 'one of those defining features of the human mind'. When you get an insight it feels like an 'aha' moment, like Archimedes jumping out of the bath yelling 'Eureka'. There is a feeling of realization, of seeing something that was there all along but you've only just been able to lay your finger on it. This is because you are engaging in a form of problem solving. All problem solving relies on working with lots of different neural networks in your brain. At the moment when certain neural and cognitive processes occur you are able to make connections that previously eluded you.

NEUROSCIENCE OF AN INSIGHT

Beeman's research tells us that certain things happen in the brain before an insight. Normally people go quiet and still just before getting an insight. In the brain there is an increase in alpha band wave activity (a form of brainwaves) just before an insight. This is over the right occipital lobe, which is the area of the brain that processes visual information. As soon as the insight occurs the alpha activity disappears.

The theory is that people are aware that they are very close to having an insight (unconsciously) and so want to shut down the visual input, giving themselves a quieter brain so they can better see the solution.

How to predispose yourself to having more insights

- Create a quiet mental and physical space for yourself. If you are in a meeting you need to become good at blocking everything else out. You could let people know you just need a couple of minutes to process, but people will soon realize this is what you do.

- Daydream – allow your mind to wander and release any attachment to where it goes, this will help you connect with the things that could lead to the insight.

- Be happy (but not too happy) – the ideal is a pleasant happiness, a little curious, open and relaxed, as opposed to being at all anxious.

- Don't try to have an insight – the more you try the more likely you are to have the same thoughts over and over again, which will lead to an impasse (getting stuck).

Why you forget

Ben asks why he, and other people, all seem to forget some things. The answer to this can get quite complex. Some of the reasons are:

- Our brains want us to forget, so work hard to do so.
- We haven't stored the information long term to start with.
- It is difficult to retrieve the information.
- High levels of cortisol are present.
- Distractions can hamper finding a memory.
- If the information isn't used regularly it can become much harder to recall.

If forgetting things regularly is a challenge, looking for patterns in what you most often struggle to remember can be useful to identify why you may be forgetting. Then you can create a plan to help you remember more effectively. If you don't know whether it is that you aren't storing it well, or recalling it easily or accessing the memory at all, it's much harder to know where to start in addressing the problem.

Improving your memory

There are some standard things that have been shown to help keep your memory in optimal condition.

MEMORY JOGGING

At Nihon Fukushi University in Japan an experiment was done to see whether jogging helped specific mental abilities. Healthy participants jogged for 30 minutes 3 times a week for 12 weeks. The control group had a sedentary lifestyle. They were all tested three times during the three months and after the last test the joggers scored nearly 30 per cent better than the non-joggers. Several other studies also show that regular aerobic exercise enhances memory. So the business case for keeping fit just keeps building!

Basically, getting the most out of your memory involves following standard advice:

- Get enough sleep – being sleep-deprived affects your memory recall and ability to create new memories (so, for example, studying when you have a newborn baby at home may be tricky).
- Spend time with friends – being social and having close friends has been linked to having a healthy memory.
- Laugh lots.
- Manage stress – if you get really stressed regularly it would be a worthwhile investment to take a serious look at your life and your mind.
- Consume a healthy diet rich in antioxidants and omega 3s.

Real or fabricated

It would be fair to say that most people believe that their memories are real. We tend to act as if what we remember is the truth. At times this can cause challenges when two people remember things differently. Ben is familiar with these types of challenges. At home, he and his wife Rebecca often have different recollections of agreements that have been made and even conversations that have been had. At work, people sometimes say they've told him things that he has no memory of; sometimes they seem to have particular expectations and Ben is completely unaware of how they got them.

CREATING FALSE MEMORIES

Elizabeth F Loftus has been responsible for conducting many experiments that examine the creation of false memories. She is both a professor of psychology and adjunct professor of law at the University of California, Irvine. One of these experiments involved participants watching a simulated automobile accident at a crossroads with a stop sign. After they had seen it half of them were given a suggestion that the sign was in fact a yield sign, the other half weren't given this suggestion. Of the group that received the suggestion, when they were later asked what traffic sign they remembered seeing, the majority said it was a yield sign. The other group were far more accurate.

FALSE MEMORIES

Iran Hyman from Western Washington University has also carried out a range of experiments in this field. One involved presenting participants with accounts of events some of which were true and some false. One of the false memories offered was that they accidentally spilled a bowl of punch over the parents of the bride at a wedding. They were then interviewed twice. During the first interview none of the participants remembered the false events, with one interviewee saying 'I have no clue. I have never heard that one before.' During the second interview 18 per cent of people remembered something about the fictitious event. The same interviewee said 'It was an outdoor wedding, and I think we were running around and knocked something over, like the punch bowl or something, and made a big mess and of course got yelled at for it.'

These experiments throw up some important realizations. It looks as if it is possible to remember things that haven't actually happened! This is contrary to how most people behave. If you were to ask people if they are sure that what they are saying actually happened they would probably think you are calling them liars. The case isn't as clear-cut as that though. If everyone realized that what they think they can remember may or may not be what actually happened, far less unconscious black-marking of people would occur.

Ben is thinking about sharing some of these studies at his next team meeting just so the people he works most with may open their eyes a little to their recollection of reality possibly not being the 'true' one. His hope is that they could all become a bit more patient and forgiving with each other.

POWER OF IMAGINATION

Loftus and some colleagues wondered if imagining things had happened would increase our belief in something having happened. They carried out an experiment, which had three stages to it. Participants were asked to rate events on a scale of 'definitely did not happen' to 'definitely did happen'.

Two weeks later they were asked to imagine that an event had happened. One example was that they were playing at home after school, heard a strange noise outside, ran, tripped and broke a glass window. During the imagining stage (which only half the participants went through) they were asked questions such as 'What did you trip on?' and 'How did you feel?' In the third stage of the experiment 24 per cent of the participants who had imagined the event happening reported an increased confidence that the event had happened, compared with 12 per cent of non-imaginers.

Further experiments have been done that go as far as creating new memories from scratch of things that never happened. What we can learn from these studies is hugely valuable. Now we know that false memories can exist we can be more sensitive to differing recollections of things such as meetings. If Ben doesn't remember something the same way that Rebecca is saying it happened, that doesn't mean she is lying or trying to manipulate him. It could simply be that they have remembered things differently.

This can happen in lots of areas of life. Take, for example, people trying to lose weight. If the measurement of this is only being done by feeling how tight clothes are after a week of effort, a person may

---

REAL IMPROVEMENT VERSUS CHANGING STARTING BLOCK

Michael Conway did an experiment with students that is linked to memory at the University of Durham. A group of students believed they were taking a 'study-skills' course that would boost their learning and recall abilities. The final exam marks were compared with those who hadn't taken the course and the study-skills group's results were worse! You could just presume that the course was a very bad one, but the students all said that the course was helpful to them. So what happened?

Before the course started the students all rated their current study skills. They also did this at the end of the course and recalled how they had rated themselves at the start. They remembered how they rated themselves initially as being a lot worse than they actually had. The likely explanation is that because they believed that they were going to get better at studying, they changed their memory of what they initially rated themselves as so that it would show an improvement compared to their new current rating.

---

think the clothes started out tighter than they actually did! In business it is useful to watch out for changing starting blocks. If things aren't well documented in meetings then there is scope for everyone potentially to work off whatever they remember.

Episodic versus semantic

There are two main types of memory. One of these types (declarative memory) is made up of two other types, episodic and semantic. Episodic memories relate to specific events and tend to be associated with emotions. Semantic memories are memories that do not record how or when we acquired the memory or any associated emotions.

Antonio Damasio, an expert on emotions (among other things), has a theory about episodic memory retrieval. He suggests that we can think of information being stored in little parcels all over the brain. The information comes together at 'convergence zones' and these are located close to the initial neurons that registered the original event. With this theory it is likely that the hippocampus acts as the central point (like a telephone exchange) where anything not required is filtered out.

This is important to Ben because he often has unrealistic expectations of how to go about storing large amounts of semantic memories, ie storing memories without linking them to experiences or emotions. Data about projects he is working on is difficult for him to remember if he just reads through it before going into meetings. He is treating the information as if he wants to hold it in his working memory, but this isn't designed for the volume of facts he is trying to hold there. Episodic memories, such as jokes the guys played on each other during the project, are easy for him to remember.

He could try a couple of different things to store the data in a way that is more effective and easier to recall. He could start by making a brief list of the things he wants to remember for the meeting and link each of them to something very familiar to him. For example:

- Company made £3.1 million profit in May = Ben proposed to his wife on the 31 May, so he pictures that day and thinks of how valuable she is to him.
- Need to save £594,000 by October = Ben pictures a Halloween cauldron with a £5.94 price sticker on it.
- Salaries are £1.63 million currently = Ben's mother was born in 63 – so he pictures his mum with a 1 on her forehead and 63 on her tummy.

Making quick links of dry data to familiar things gives the brain more chance of being able to recall the information.

Brain areas

SWIMMING RATS

An experiment with rats supports the idea that the hippocampus is important in the memory system. In case you aren't familiar with rats, they can swim but it isn't their favourite pastime and so they will try to get out of water as soon as they can. When you put them in a tank filled with water with a platform (out of the water) at one end they will swim to it. Once the rats have learnt where the platform is, the experimenters submerge it so the rats can no longer see it. They now have to remember where it is to get out of the water. Rats with lesions to their hippocampus find this tricky.

HIPPOCAMPUS MAPS

A study in the 1970s by John O'Keefe at University College London first gave scientists an insight into how the hippocampus worked. O'Keefe, along with colleague Lynn Nadel, proposed that the hippocampus forms a spatial representation of the world that is independent from sensory inputs. They explain that one of the functions behind this is to create a context for memories to connect into (O'Keefe and Nadel, 1978).

Working memory

In Chapter 10 Kate experiences the limitations of her working memory. Ben is finding he has similar challenges. Stuart explains how the working memory works to Ben, in the same way he did to Kate, and asks whether he thinks there is a better way to organize himself for meetings. Ben reflects and decides that if he had one sheet with everything that he would be expected to lay his hands on in terms of data, that would be useful. Then he could have supporting documents that would take him time to go through to elicit information and he would just have to say that it would take him a moment to find that answer. For really important meetings he could use the trick of associating facts and figures with familiar things, but most of the time he realizes he probably isn't expected to store them all in his head.

The ideal situation is when everyone understands that in meetings your working memory is best placed to process things and hold things very short term for you. Trying to hold lots of facts and figures is tricky and not the best use of your resources.

Emotions and memory

Emotions and memory often go hand in hand. Emotions affect the encoding phase of forming memories and the consolidation phase. The things that you are able to remember most vividly are often those with strong emotional connections. If you want to remember something it is useful to link it to an experience or an emotion.

When working with other people it can be handy to create shared emotional experiences to help them remember things. It will also bond the group together. Emotion can also skew what people remember, producing both good and inconvenient results.

Unconscious memories

One of the things that Stuart knows is challenging Ben is the general unpredictability of the people he works with. Ben has said that he thinks they are illogical, irrational and this makes it hard to work with them. If he could just put his head down and get the work done there wouldn't be any problems. He would be much more productive. It's the dealing with people that can drain his energy, motivation and patience.

It is not an unusual belief to hold that people aren't logical. It is a belief that Stuart disagrees with though. Most people expect others to behave based on the information that they have. They forget that the individual will have consciously far more information, and unconsciously about 80 per cent more information. So to conclude that someone is illogical or irrational on that basis is missing a trick.

At any point of interaction with another person there are a huge range of possible things that may be going on for them. Here are some examples:

- Their emotional state may not be conducive to the conversation you want to have with them.
- They may be distracted, have an overworked prefrontal cortex or lacking enough dopamine to focus.

- They could simply be primed to respond in a way you hadn't considered.

Priming

You'll remember priming from Chapter 8 with Jessie as 'the Neuroscientist's best kept secret'. Priming activates certain neural circuits and sets us up to respond in a certain way. It is a form of implicit memory, which is the type of memory that helps us do things without us having conscious awareness of what is helping us. Priming is a form of subconscious preparation. It can be used in ways that really help you out, enrich your life, and increase your productivity, for example:

- You wear a particular suit for meetings where you want to be in top form (unconsciously you know you have been successful wearing that suit many times before).

- You talk about a respected role model with your colleague on the way into work (unconsciously aligning your values, attitudes and beliefs with the role models and subsequently programming you to behave more like the role model).

- You look at a photograph of a favourite holiday with your partner as you walk into the house (unconsciously associating with a state of relaxation, connection and love).

Priming can also cause a lot of challenges and for many people who don't know that it is going on this can compound the problems. For example:

- Your managers are being short and dismissive of you (they have just read an e-mail from their boss saying they need to be firmer with the team).

- You get a 'no' to a charity request from someone when you were expecting a yes (the person's intention for the week was to be focused on a particular report and to not get distracted).

- You are sidelined for a big exciting project (the decision maker had a conversation with someone about how you cope with the less high-profile work).

There are things you can do to almost counteract priming, and you can think of that as offering a new prime to the person. So in the examples above you could do the following:

- Next time you need something from your managers explain clearly what you need from them, pre-frame that you have two questions and that it should take four minutes of their time to answer you.

- Leave it a week and then ask to schedule an appointment. Send an e-mail with a breakdown of the previous discussions around charity involvement. Also list the benefits to this situation and make it clear that you will take responsibility for things.

- There's not much you could do with the last one in the list, as you almost certainly won't know why you were sidelined.

MYBW top tips for remembering stuff

- Decide how you need to remember it – does it need to be in your head, or can you strategically use prompts?

- Know how and when you want to recall it.

- Be prepared for others remembering things differently from you (confirm things in writing if it is important to be on the same page).

- Understand that people don't always realize their unconscious is affecting their memory.

MYBW top benefits for understanding the process of remembering stuff

- Use your mental capacity efficiently and make empowered choices.
- Work from an enlightened place with other people.
- Support teams to remember and forget to maximize results.

Part 3
YOUR COMPANY

The third key to this puzzle of creating life to be as pleasurable as possible is the company that you work in. If you have mastered your personal efficiency, effectiveness and productivity and understand how to best work with your colleagues and clients your attention turns to your company. You have power to affect your organization.

Companies of the future will be enjoyable to work in. People being unhappy for a large part of their working life is a dying trend. Individuals' expectations are rightly growing and the companies that pay attention and grow too will be the ones that survive.

The companies that get the best out of their employees and work towards a big vision best understand what employees and customers need. These are the organizations we should be learning from. Most of the companies discussed here have an entrepreneurial energy to them. While this in itself isn't a necessity, it happens to be where more examples of great practice are in alignment with what neuroscience teaches us is important for success.

All companies are capable of change. Throughout history we have seen huge 'old-school' companies embrace new ways of working and be more successful as a result. Neuroscience confirms that these changes are possible and enlightens us as to how best to make the transition.

13
Leading with your brain switched on

The model that brings cutting-edge neuroscience into the world of leadership

Kate feels that she is going to have to leave the organization she is working for. She has always thought it important to respect the people leading you and she realizes that she just doesn't. This has been building up over such a long period of time she isn't even 100 per cent sure what all the component problems are, she does have an overall feeling of disappointment in the company though.

During a recent meeting with her boss Sue, Kate started to feel really uneasy. She could not put her finger on it though. Sue was talking about next year and how the company is going to go from strength to strength as they build on the success expected at the end of this year. Kate's mind was wandering to try to distract herself.

Sue seems to be very good at talking generally about things but Kate explains that she often finds it difficult to know what her specific expectations are. Kate knows what she wants to do but in the past she has presumed that Sue would be okay with things and this has turned out not to be the case.

It isn't even just her immediate leader Sue who is challenging Kate. Overall, she just isn't feeling as if the company is congruent any more with what she thought the leadership team stood for, or that the vision is being lived out. The teams don't seem as valued as they used to and people don't seem to be consulted as much. She doesn't feel as if the leaders put as much into this company as she does and that is making her question if she is in the right place.

The people in the most senior leadership team have often said they will do things or take a certain direction and then nothing has happened. Kate recognizes that the direction has to change sometimes but she wished they would communicate this more effectively.

This chapter is about leading people in a way that connects with how their brain actually will respond to reward both leader and followers to enable a whole company to be more successful. The bonus is that everyone is better placed for dealing with change.

Leaders today

Leaders today have the opportunity, even a responsibility, to understand how their brain works. Leadership is directly linked to people. Our knowledge of how people work has moved on substantially in the last 20 years. Not making use of this knowledge is putting you at a disadvantage. The old style of leadership development work involved a lot of guess work. Now we have access to reams of research that remove a lot of the hit-and-miss approach of the past.

LEADERSHIP AND MANAGEMENT

It is amazing how often these terms get used interchangeably when they are in fact two very separate skill sets. According to Peter Drucker and Warren Bennis (quoted in Covey, 1989), 'Management is doing things right; leadership is doing the right things.' Covey (1989) says, 'Management is efficiency in climbing the ladder of success; leadership determines whether the ladder is leaning against the right wall.'

The synaptic circle

Stuart introduces Kate to a comprehensive neuroscience of leadership model, called the synaptic circle. This is one of many Synaptic Potential use with companies all around the world. It covers the fundamentals a leader needs to be aware of in order to best work with people, taking into account neuroscientific insights. There are three components to the model and Stuart focuses on the second component with Kate as this will make the biggest impact for her at this time.

The second component of the model is a series of vital elements that all begin with the letter 'c'. They can all be applied to:

- yourself;
- your clients and colleagues;
- your company.

The six elements are confidence, certainty, celebration, control, connection and contributution. They all earned their place in this much loved model because there is strong research saying that they are critical to success. Each has neuroscience underpinnings. Not all of them are as they first seem so Stuart uses this coaching session to elaborate on them and see what Kate thinks.

Confidence

Confidence is vital on several levels:

- Leaders themselves need to be confident.
- Colleagues and clients need to have confidence in the leaders.
- Everyone needs to have confidence in the company.

So what actually is confidence? It is generally thought of as a state of certainty. There are lots of ways a person can convey this internal certainty. In fact, many personal development programmes teach ways to raise confidence. The risk is that people try to portray confidence externally when the internal certainty isn't present. This is a bad plan and we will look at why true, real confidence is best, and if that isn't achievable then an honest lack of confidence is better.

Threat response

Many of the synaptic circle elements evoke either a neurological threat or reward response depending on how well they are fulfilled. These responses are hugely important because they have a direct effect on the efficiency, effectiveness and productivity of people. If employees lose confidence in their leader it can directly affect their productivity, which ultimately costs the company money. So ensuring that leaders genuinely know how to work with confidence is vital.

OVERVIEW OF THE NEUROSCIENCE OF THE THREAT RESPONSE

When people experience a threat response a series of events inside their bodies are triggered. The process itself uses up oxygen and glucose. This has the effect of decreasing working memory capacity. You need your working memory for creative insight, analytic thinking, problem solving and even just simply holding something in your very short-term memory. These are all impaired when you are having a threat response.

A threat response also affects the amygdala, anterior cingulated cortex and frontal lobe. The release of cortisol is triggered and this results in decreasing immunity, impairing learning and affecting memory. All together this spells disaster for efficiency, effectiveness and productivity.

The brain cannot distinguish between real and unreal threats. If you see what may be a snake near you on a country walk it is safer for you to respond as if it is one, rather than presume it isn't. Your limbic system responds quickly to keep you safe. If it turns out that there isn't a snake your brain usually calms the threat response down. In organizations things are normally far more complicated and your brain often struggles to determine if something is a real threat, so it keeps you safe by presuming it is.

When a threat is perceived an important system called the hypothalamic-pituitary-adrenal (HPA) axis kicks into action. This system comprises interactions between the three named brain areas. The HPA system triggers the production and release of cortisol. This prepares the body for action; the blood pressure and heart rate increase, the lungs take in more oxygen by increasing the breathing rate and blood flow can dramatically increase.

Neurotransmitters are also released as a result of the HPA system's activation. These chemical messengers activate the amygdala, which then triggers the brain's response to emotions. The chemical messengers also tell the hippocampus to create a record of this emotional experience in the long-term memory. Finally, the neurotransmitters also suppress activity in the frontal lobe, which means that short-term memory, concentration, inhibition and rational thinking all take a nosedive. Trying to handle social interactions elegantly or do cognitive tasks becomes very challenging while in this state.

Understanding status

A common challenge in companies is people not understanding the power of status. People are programmed to respond to perceived status. If they think that someone more powerful, important, beautiful, intelligent (the list could go on) is present then it is normal to have a minor threat response to that. If we have just done something that boosts our status (or even if we just believe it boosts our status), such as making a great comment in a meeting or securing some new business, we have a reward response.

When working with people you want to get the most from it makes sense to reduce their threat response to you. Stuart explains to Kate that by 'status levelling', which involves either reducing your status or elevating their status, you can help people to be more productive. For example, when Kate praises the people she works with they get a little reward response. Their dopamine, serotonin and testosterone levels increase and their cortisol levels decrease.

Some organizations have a culture in place that means that most members of the workforce feel they have a low status most of the time. Highly competitive environments, where the top 10 per cent of performers are celebrated and the rest are looked down upon, will tend to have challenges.

Mirror neurons

Integrity is something that a lot of people cite as important. However, when it comes to confidence there is a whole school of thought that promotes 'fake it until you make it'. This implies that pretending to be confident is a good plan until you actually are confident. This strategy may potentially raise your confidence (because part of confidence is your internal dialogue, and if you are standing confidently, speaking confidently and looking confident, you are more likely to tell yourself you are confident). However, it may also cause other problems.

If you lack internal congruence it is likely that other people will pick up on this, subconsciously or possibly even consciously. A subconscious awareness that someone isn't congruent is enough to set alarm bells ringing and for people's threat response to be activated. If you want to evoke clean responses in people then it is better to be honest.

Sue isn't confident about the organization achieving their goals for the end of this year. This will scupper all their plans for next year. When she talks about it she thinks that she needs to put on a brave face for those in the team. She thinks that they will work better if they believe everything is fine. The problem is that unconsciously they are picking up on incongruence, which makes it harder for them to trust their leader.

Company confidence

It isn't just individuals that we can have or lose confidence in. What a company stands for and how congruent that is can be hugely powerful. Clients and customers can receive mixed messages or have their threat response evoked, which makes them far less comfortable buying from you and/or working with you.

SIX SENSES

Sonu Shivdasani and Eva Malmstrom are the married couple behind the Six Senses brand. They created Soneva Fushi, which is a luxury resort in the Maldives. This is no ordinary resort though. Through its creation the couple have redefined what is possible, and perhaps even expected, by luxury resorts. They were so confident in what they believed in and what they wanted to create that they didn't bow to common convention.

Here, the creation of something new defined them and there was integrity across the board. This is very important. At Soneva Fushi there is an organic garden for fresh and organic food. Guests are given a hessian bag to take any plastic rubbish home with them. The guests don't wear shoes; so neither do the executive group. They are currently working on becoming not just carbon neutral, but carbon zero.

The guests who visit know what to expect and so do new employees. People don't just work with Six Senses for their pay cheque; they 'aspire to a lifestyle and embrace our philosophy'. It wouldn't fit if employees were given restrictive rules that didn't honour their natural abilities. The Chief Talent Officer says that instead 'We tell our hosts what Asian hospitality means and what our philosophy of caring for the guest is all about and then we leave it up to them.'

Who Sonu Shivdasani and Eva Malmstrom are is clear for everyone to see and experience. This enables Soneva Fushi as a company to be very confident. It also enables employees to be confident in the leadership of the organization, their role in it and their higher purpose from a career perspective.

STAFF

It seems like a small thing, but many companies are starting to call their staff things other than staff. Disney has casts rather than employees. Zappos has Zapponians.

Kate is not alone in feeling frustrated about her leaders. She feels she is putting everything into her role. She works long hours, thinks about work at night and on holidays, talks to her friends about the company and it often feels as if it comes first in her life. It is a huge part of who she is. Her role at the company is a big part of how she sees herself overall in life. Stuart asks how she feels about this as her frustration seems to be building. Kate exclaims that she is happy that she is as dedicated as she is. She enjoys being this way and wants to be someone who gives her all to any company she works for.

The problem, she starts to realize, is that she doesn't have confidence in her leaders to have that same level of dedication. She knows logically that she can't decide for her leaders how they spend their time or what they think on holiday, but it makes her question herself. If they aren't prepared to put in at least as much as she is, what does that mean? Is she stupid for believing in this company as much as she does?

ZAPPOS – TONY'S CONFIDENCE

The story of the early days of Zappos, a major online retailer, is one of pure determination. The leadership demonstrated from the core team members involved was admirable. Started by a guy called Nick, with no shoe-selling experience, who found a guy called Fred, who did have some experience with shoes, they approached Tony Hsieh and his investment company for some funding. When their first lot of funding ran out Tony and his business partner Alfred invested again (which was against their original business plan, but it was either this or let Zappos go bust). Tony also invited them to move into his loft as a temporary workspace. Soon Tony became a full-time employee of Zappos too and had taken on a leadership role.

When the remainder of their investment money had been ploughed into Zappos Tony began taking a little out of his personal account every few months to keep the company afloat. When they had to ask employees to take pay cuts Tony opened up three lofts he owned to house employees without charging them rent.

When Tony's personal cash reserves were dwindling he sold the real estate he owned and funnelled the proceeds back into Zappos. Eventually he had sold all but his own home and a loft which, due to the poor economy, there were no interested buyers for. Eventually the need for cash meant Tony did find a buyer for the remaining loft – for 40 per cent less than he bought it for.

Alfred said 'As your friend and financial adviser, I'm advising you not to do it. It might pay off in the long run, but it's not worth the risk of being completely broke'. His parents weren't thrilled and asked if he was sure he wanted to give up all the money. Tony's thoughts however focused on the faith of his team mates. Fred had given up a great career to work with Zappos. He had a new house and children to take care of. Tony implied that if Fred could risk it all, then he should be willing to too.

There are many different ways for leaders to show their commitment and confidence to their organization. They range from the simply stated, but tricky to implement consistently, 'character of integrity' to the extreme example of Tony Heish. The example of Zappos is a good one because it shows the confidence of its leaders in their company. They knew it was risky at times, but they were committed to doing everything they could do to give it the best chance at succeeding and being all it could be.

Imagine working for a company where you know your CEO has gambled millions of his own money, gone through countless sleepless nights and made choices most people would scorn him for. His cards are very much on the table. He has proven how he feels about the company.

Other great leaders have elicited equal confidence in those who work with them through very different means. Scientists have shown unwavering commitment to a particular discovery. They have endured professional snubbing, the withdrawal of funding, public humiliation and yet remained dedicated to a particular line of investigation.

Certainty

When Kate feels uncertain about what is happening she has a threat response. Although she is often unaware that this is what is happening. Stuart explains that the same thing happens when Kate feels that Sue isn't explaining her expectations clearly. Kate feels uncertain about where she is aiming, what the boundaries are in terms of how she gets there and how she'll know she is doing a good job.

At times this can be really useful because it causes you to pay attention, be alert, be able to respond quickly, etc. However, most of the time a lack of certainty isn't deliberate or conscious and has negative effects. There are things that Kate can do to take control of times where she feels there is an absence of certainty.

The risk with increasing Kate's awareness of what makes a great leader is that she would simply judge others as poor, rendering her helpless in the grips of others' terrible leadership. This is where Stuart's role as her coach is vital to ensuring that she has coping strategies that will enable her to be led effectively even when others' leadership isn't up to scratch. She won't always be able to create an optimal situation, but should always be able to ensure that she is able to be productive and happy. That's not withstanding the valid option to leave a company. Sometimes that is the wisest thing to do.

The strategies will vary depending on how the uncertainty presents itself. In the current situation she needs to take a two-pronged approach. First, she needs to work with Sue to help her understand what she needs in order to best do her job, and subsequently to help Sue shine in her job. She can experiment with giving Sue written lists of things she needs or having meetings with her and verbally trying to elicit answers from her. The second thing Kate needs to work on is to create her own certainty. In this instance that may mean creating her own answers to her questions. Obviously this could cause problems if Sue eventually says the answers don't match hers. The way around that could be to send Sue documents that detail the basis on which Kate is moving forward, outlining the assumptions she is making in the absence of any confirmation. She could give Sue a short amount of time to correct any of these assumptions explaining that she will move forward as if they are true if she doesn't hear back within that time frame.

Company certainty

Companies as a whole are also able to give people certainty. They can assure the public of who they are, making their customers feel good about buying from them. They can also make their employees feel certain – about whom they work for, what their role in the bigger picture is and how they are going to work.

INNOCENT

innocent is a clean company. It started when three 26-year-old guys wanted something healthy, natural and delicious that they could drink every day. Richard Reed, one of the three, realizes that 'what has created a huge amount of value in innocent has been our congruency'. Congruency gives people certainty.

Right from the start when everywhere was turning them down for financing the entrepreneurs allowed who they were to benefit them. Their young, energetic, almost jokey approach of sending an e-mail to people they knew asking 'Does anyone know anyone rich?' paid off.

Immediately they ran into challenges from other people who wanted to advise them on how to be most profitable, avoid wastage and give long shelf lives. They resisted the temptation to go down that route, and realized that they needed to be congruent. They were called innocent. Their business idea was fresh and natural. Their product had to be better than others or they didn't have a business idea.

If you've ever bought an innocent smoothie you'll have been hard pressed to not take notice of the side of the carton. They're brilliant; written in the guys' sense of humour and in language the consumer can relate to. It is another example of innocent communicating who the company is, giving another bit of certainty, helping us categorize it in our mind most effectively.

The company's level of certainty and congruency doesn't obviously translate into profits. That isn't to say that it isn't profitable, just that it can be hard to measure. Innocent believes in getting its fruit from farms that have high social and environmental standards... this means its pineapples cost 30 per cent more than normal! The company also gives 10 per cent of profits to charity.

Celebration

In this context we are using celebration to mean anything that triggers the neurological reward system into action. This can happen many times in a day for some people at work, and for others be much more scarce.

NEUROSCIENCE OF CELEBRATION

When something triggers the reward response there are several areas of the brain that are activated. The last area in the series of obscure brain areas to be activated is the primary release site for the neurotransmitter dopamine.

Dopamine is the star of the show. It is heavily involved in reinforcing behaviours. People feel good when dopamine is flooding their system and they remember this and want to do whatever triggered the release again.

Dopamine also has positive effects on cognition, memory, attention and problem solving – all great things for leaders to induce in their teams!

Company celebration

Celebrating things seems to be something that lots of companies have seen value in. Here are some of the ranges of things that are being done. Celebrating good service is done in a variety of ways. Virgin gives great service awards where customers have nominated staff. They celebrate by taking the employee to the United States and they get to have dinner with Richard Branson. Pret A Manger (a UK-based chain of sandwich shops) team members get a Tiffany Silver Star sent to them if they get mentioned by name by a customer. Richer Sounds give their winners of richer league (the service-based measurement scheme) a Rolls Royce or Bentley for the weekend.

Some forms of celebration are in the style of general rewards. Virgin have something called a Virgin Tribe card, which enables employees to receive discounts at all virgin companies. Other companies like to create a culture of celebration generally. Carphone Warehouse have a beer bus once a month for their employees.

Other forms of celebration require some pretty high levels of loyalty. John Lewis, which is the UK's largest example of worker co-ownership, does something quite special for its partners. After 25 years' service they get a six-month fully paid sabbatical.

Control

When you feel as if you don't have control over something you are likely to experience the threat response, as we saw in Chapter 9 with the addicted rats. As a leader it is best to facilitate people to be autonomous. At the very least leaders need to find strategic ways of making people believe they do have a certain amount of control. Companies do this in all sorts of inspiring different ways.

Banyan tree (world-wide luxury resorts) gives staff creative control with a framework. So when they are decorating a bed in a beautiful resort they must leave the standard gift for the guests but can also leave a flower or anything else they choose to as well. At Pret A Manger there is an experience day for new recruits. At the end of this day the team votes on whether or not to hire them. This is quite unusual but gives the team a huge amount of control over the people they will be working with. It gives them ownership and is likely to make them invest more in making the new recruit work out. At Pret A Manger the staff are told 'to greet customers when they arrive, look them in the eye when they put money in your hand and say something when they leave... and be yourself'. This enables staff to engage their brain, if they choose to, with each and every customer. They could be observant, really connect and make a much better impression than is possible with a standard company-issued line that they have to say. Tesco say that they encourage people to bring their brains to work and to have an entrepreneurial spirit. Finally, First Direct have a package called PLUS where people can tailor their benefits to their stage in life. This creates a great sense of autonomy.

There are lots of ways that companies restrict employee autonomy. One of the ones that make least sense is this. Many customer call centres have scripts that the employees are supposed to follow. This reduces a person's autonomy greatly. It also misses out on the opportunity of allowing individuals to use their brains to their full capacity.

Connection

In 2008 John T Cacioppo and William Patrick showed that loneliness is itself a threat response to a lack of social contact. The same neurochemicals that flood your body when you are in physical pain are activated. Being isolated, ostracized or ignored is a productivity kiss of death. On the other hand, when you are connecting with people you release oxytocin and this helps reduce blood pressure and cortisol levels. It also helps with positive social interaction.

Connecting people in an organization is a relatively easy thing to do. The ways you can do it depend on the specific organization and what will work best for you. Virgin chooses to get all the HR people together every three months. Similarly, the IT team get together and it's the same for marketing and finance staff. Pret A Manger has a tradition of every Friday night taking over a pub in London with an open invitation to all team members. At its Christmas and summer parties it has over 2,500 people attending. Then five times a year head office is cleared and everyone goes to work in the shops. This is great for connecting people in more senior positions with customer-facing team members. Everyone in head office also has a buddy shop, where they are connected to the customer-facing team, rather than it feeling as if the leaders are secured in an ivory tower making strategic decisions.

Umbqua Bank is well known for its random acts of kindness. This is a way of connecting, normally in an emotional way, with its clients or potential clients. Members of the bank will take an ice cream truck out into the street and give away free ice cream to passers by. They will go into a coffee shop or restaurant, choose a table and pay for everything that people who sit at that table order for the day. When the bank opens a new branch you may find as a local resident a pot plant delivered to your door with an invite to pop into the new branch for a bag of free Umpqua Blend coffee. The company aims to connect with people in its local area and it is very proactive about how it does this.

Zappos recognizes that great business often happens when people are in flow. They are relaxed, you can connect on a deeper level and conversations happen naturally. On the last Friday of each month Zappos organizes a golf tournament where vendors are invited to play. A conversation with one of the reps on the course resulted in the company getting into eyewear, and its eyewear offering is now one of the largest online.

Zappos has also been consistently impressive at connecting internally with everyone. Tony sends out powerful e-mails that are transparent, clear and often from the heart. These have a big effect on enabling people to feel part of a pretty special team. One e-mail, sent on the 11 November 2008, finished with 'Remember, this is not my company, and this is not our investors' company. This company is all of ours, and it's up to all of us where we go from here. The power lies in each and every one of us to move forward and come out as a team stronger than we've ever been in the history of the company. Let's show the world what Zappos is capable of.'

Contribution

Contributing makes you feel good. Experientially we all know this; however, now we have the neuroscience to back it up. The mesolimbic reward system (which you met earlier on) is engaged by making donations to charity in the same way that occurs when you receive monetary rewards. Corporate social responsibility is viewed in many different ways by organizations today.

VENEZIANA PIZZA

One of my favourite pizzas from Pizza Express is called the Veneziana. Every time someone buys this yummy pizza 25p is donated to the Veneziana Fund, which preserves and repairs buildings in the UK originally constructed before 1750. Over the years they have supported many different charities in this way. Local branches are also involved in charity and education schemes.

Richer Sounds give 7 per cent of pre-tax profits to charitable projects and allows the staff actively to work on projects on paid leave. The company also invests 1 per cent of profits in a hardship fund that can be used by employees if they have problems.

MYBW top tips for neuroscience-based leadership

- Check any problem or challenge against the synaptic circle of confidence, certainty, celebration, control, connection and contribution and look for what is missing.
- Be strategic and systemize each of the elements into your personal leadership of yourself.
- Work with other leaders to implement systems to integrate the synaptic circle into the whole company.

MYBW top benefits for neuroscience based leadership

- Individuals are more successful, fulfilled and loyal within companies.
- Organizations are better placed than individuals for dealing with change.
- People are demanding more from companies they buy from. This best equips you to deliver.

14
A culture of more than psychological safety

Today Jessie is a bit frustrated at herself. She has had an important meeting with a senior decision maker from the local council. If it had it gone well, they would have secured more funding and been able to help more people. But for some reason it didn't. Jessie couldn't quite work out why. Their offering was perfect and would save the council money – it was a no-brainer!

Stuart knew Jessie well by this point and believed she was excellent in her role. He wanted to explore whether something else might be going on, that perhaps Jessie wasn't consciously aware of.

This chapter explores psychological safety and the impact it has on people. It also goes beyond this concept to challenge the opportunities for all cultures to be intentional. The highly influential role a culture plays in performance means it is an essential to all organizations.

A strong organization requires a culture of psychological safety to facilitate the expression of skills.

WHAT IS PSYCHOLOGICAL SAFETY?

Some describe it as an environment where people feel safe enough to take interpersonal risks. This may be through speaking up about their feelings, thoughts or observations. It may be being able to ask questions.

MIT professors Warren Bennis and Edgar Schein suggested back in 1965 that psychological safety was essential for people to feel secure. It was important in order for people to change their behaviours. When people are psychologically safe, they can focus on collective goals and problem prevention rather than on self-protection. This is huge.

In 1990 William Kahn looked at how psychological safety enables personal engagement at work. He found that people were more willing to express themselves physically, cognitively and emotionally when they were psychologically safe. The other option being to disengage, to withdraw and defend themselves. He also found people believed that they would be given the benefit of the doubt more. Essentially they would be trusted and respected.

Without it you risk:

- Employee satisfaction
- Sharing of information
- Asking for help
- Experimentation
- Innovation

Jessie has never thought that much about culture before. At medical school the way some consultants treated you was just what you came to expect. Same with some of the nurses. It was normal to be on a ward round and to be put on the spot to answer questions. If you got it wrong you might be made fun of. It was embarrassing. Some of her friends dreaded those rounds because they knew they'd feel ignorant or incompetent. You just hoped you didn't get picked on too much that day.

There were some nurses Jessie didn't ask for help from as a junior doctor. Initially, when she moved to a new ward she'd try to work with everyone. But it soon transpired that some nurses loved helping out the attractive male doctors, but seemed awfully busy when she needed help. Of course many of the other nurses were absolute stars. It was just easier to ask those who were open to helping rather than deal with the awkward rejection.

What happened there actually happens lots of places in different ways. It is potentially really damaging. The challenge of not feeling able to speak freely to everyone we work with needs addressing.

THE TRAINING ROOM

One of the most successful role-plays I've witnessed was during an event with a bank. They split up all their client development people into teams in different rooms around a hotel. Each team was then visited in turn by four different characters. One was the CEO of a fund, another the CFO of the company, there was another employee and finally the CEO of the whole company.

Watching these highly successful, highly intelligent individuals question the different characters was fascinating. They started really well. The CFO and CEO of the fund were very amenable. They were friendly and clear and warm in their communication. This wasn't obvious until we saw the contrast frame of the CEO for the company.

This character was standoffish. Irritable almost. He challenged 'silly' questions. He told people they should have done their research before the meeting. Said that's what Google is for. Almost immediately the questions dried up. The fast-paced, engaged, curious intrigue that had powered the information gathering of previous interviews was brought to a crashing halt.

The additional factor was that in real life the man playing this character was senior to everyone else in the room. What he thought of these people mattered.

Anyone who asked a question from then on phrased it carefully and was far more formal than before. They elicited the least amount of information from this character. The feedback was that this wasn't a good interview.

Creating this dynamic wasn't the purpose of the role-play but for people to have this experience and be able to reflect on it was hugely valuable.

Stuart shared this story of the bank with Jessie and asked for her reflections. Jessie exclaimed that it is obvious. Of course the CEO was in the wrong. He wasn't creating an environment where the people could think most clearly and speak most openly so they could get the best information to serve him. When pressed Jessie could hypothesize some of the science that would sit behind how this scenario negatively impacted people.

SCIENCE BEHIND PSYCHOLOGICAL SAFETY

If we take this concept of safety to be the opposite of at risk of harm then we have a lot of neuroscience that can explain why it is so critical.

We summarize this response as the threat response. Someone doesn't have to feel physically threatened to have the cascade of events occur in the brain. A person may consciously experience feelings of fear, but these too may be below the level of awareness. However behind the closed hemispheres of the brain the amygdalae are activated. The networks involved in a fear response are energy intensive, and the resources are diverted away from the prefrontal cortex. Working memory capacity has been shown to decrease and this impairs analytic thinking, creative insight and problem solving ability.

There are many things that can trigger this response.

Then the ball dropped for Jessie. This is what had happened in her meeting this week. She knew this guy from the council was senior, and he dressed in a smart suit. He didn't do anything to put her at ease. She hadn't felt scared in the meeting. She would rationalize that while he was very accomplished in his field, she had been to medical school and shouldn't feel intimidated by him. Consciously this was her experience. However, she recognized something was going on at a deeper level because she just didn't ask the quality of question she was happy with. She didn't challenge him the way she had planned to. Nor did she seem her normal quick thinking self when new information was shared with her.

But what could she do about him? Sure, it would have made more sense if he had created an environment where she felt more relaxed

---

WORKING MEMORY

Your working memory can be thought of as your very short-term storage facility. It is incredibly useful but has some usability points that most people aren't aware of and can really cause you challenges. It enables you to hold small amounts of information in your head typically in order to do something. It might be to remember directions on how to get to a meeting. Or a set of numbers someone has just given you so you can calculate some figures that are important for everyone in a meeting to know about. It might even be several opinions that need to be balanced and taken into account to decide on a new direction.

The things you need to be aware of with your working memory are:

- It has limited capacity.
- If you get distracted you'll most likely forget what you were holding there.
- If your prefrontal cortex is overworked you are likely to struggle to remember things.
- Stress can reduce your ability to hold things in working memory.

---

and able to think clearly. He would have got more from her and they would have had a better meeting. But Jessie didn't control him. Only herself.

Jessie was thinking about the organization that she was building. It had never occurred to her to be like some of her old consultants. She'd always just thought that was because it wasn't who she was, but upon reflection she realized it was more conscious than that. She knew it wasn't the best way to create a good team. It didn't help people perform at their best. She wanted to check in with her team though. Make sure there wasn't anything she or anyone else was doing inadvertently that was creating a less safe brain environment for them all.

Jessie presumed she would score pretty well on these questions. Stuart took a little while to explore what it would mean if she didn't. At first this idea was hard for Jessie to entertain, but in time

TEMPERATURE CHECK: MEASURING PSYCHOLOGICAL SAFETY

People respond on a scale of 1–7 from 'Strongly disagree' agree to 'Strongly agree'. (Some of the questions are positively expressed while some negatively, this is good data gathering practice).

1. If you make a mistake on this team, it is often held against you.

2. Members of this team are able to bring up problems and tough issues.

3. People on this team sometimes reject others for being different.

4. It is safe to take a risk on this team.

5. It is difficult to ask other members of this team for help.

6. No one on this team would deliberately act in a way that undermines my efforts.

7. Working with members of this team, my unique skills and talents are valued and utilized.

Amy Edmondson used these seven survey items in her dissertation[18] and have since been widely used by other researchers in this field. Check out her book *The Fearless Organization.*[19]

she, at least consciously, concluded that it would be an opportunity for her to learn. It would probably hurt a bit too though. She didn't want to think that anyone wasn't happy or anyone felt they couldn't speak up.

Stuart wanted Jessie to mull over these ideas and let less conscious networks in her brain get to work to bring more things to Jessies attention. So he suggested a walk in a nearby green area. This wasn't the first time they had done this so Jessie was fairly relaxed about it.

As they walked Jessie started to question whether her team really did feel able to open up. She recalled that her graphic designer would always work from Jessie's ideas. By suggesting the idea she didn't mean it had to be that. She wanted her to be able to challenge her ideas. To tell her there was a much better way if that's what she thought. But she realized she never had. That may be because

Jessie's idea was always amazing, but it may be because Steph, her graphical guru, didn't feel able to voice things.

It started to dawn on Jessie that this may be happening with several team members. She can't easily see when people don't feel able to speak up. But – she was a good person, she was nice, was it really possible people didn't feel comfortable with her? – she quizzed Stuart.

VW'S CHALLENGE

Bob Lutz suggests that powerful former CEO of VW (Ferdinand Piëch) is more than likely the root cause of the diesel-emissions scandal. In case you missed it this scandal was big. Clean award winning diesel engines were not what they seemed. And it was hard to believe that no-one knew about this. Prosecutors went on to identify more than 40 people who were involved. The CEO at the time, Martin Winterkorn, resigned taking full responsibility while denying wrongdoing.

When Bob praised the then new Golf, Piëch is said to have responded: 'I'll give you the recipe. I called all the body engineers, stamping people, manufacturing and executives into my conference room. And I said, "I am tired of all these lousy body fits. You have six weeks to achieve world-class body fits. I have all your names. If we do not have good body fits in six weeks, I will replace all of you. Thank you for your time today."'[20]

The suggestion is that he took a brutal approach and the culture was one of fear and intimidation. The risk of losing your job today for sure or at some point in the future maybe for most people is easy. They temporally discount and chose to do what it takes to keep the job today.

No one is suggesting Jessie's style is anything like Piëch's or Winterkorn's. There is no evidence her team are scared of her or that they feel their jobs are at risk at any time. However, it might still be possible that there is the opportunity to improve their performance by improving the culture.

Unfortunately, creating a less psychologically safe environment is possible unintentionally. This will be the stretch for all the great leaders and managers out there. How can we intentionally create an even safer environment for our people's brains to really perform at their best?

FEELING CONNECTED

Naomi Eisenberg from UCLA conducted a fascinating study[21] that involved participants playing a computer videogame called cyberball. Each participant was told that they were going to play against two other people in a three-way simple throwing and catching game. In fact each participant played against the computer. They saw on the screen two other little characters, along with their own character's hands, and still believed that other participants were controlling these other two characters.

Initially the three characters threw the ball to each other, playing a nice game involving everyone. Soon the human participant was receiving the ball less often and eventually was left out completely. They believed that the two other people playing (it was in fact just the computer) were now throwing it exclusively to each other. They later reported feeling distressed.

During the whole process they were in an fMRI scanner, which enabled the researchers to see what was happening in their brains. It showed that the anterior cingulated cortex (ACC) and right ventral prefrontal cortex (RVPFC) were active during their exclusion from the game, when they felt distressed. They were able to conclude that the RVPFC regulates the distress of social exclusion by disrupting ACC activity. This is useful to know because we know several ways the PFC can become ineffective in its roles and this could predispose someone to increased or prolonged suffering from a social perspective.

Previous studies have highlighted the brain areas that are activated during physical pain. The researchers were therefore able to conclude that physical and social pain share a common neuroanatomical basis. This means that we are strongly alerted to any situation where we have sustained damage to our social connections. We are then in a position to do something to restore them.

The old saying 'Sticks and stones may break my bones but words will never hurt me' doesn't have any neuroscience grounding. In fact what we see is the exact opposite. In meetings if a person feels left out they may proactively try to involve themselves or they may retreat inwardly feeling hurt and worried. Being aware of this gives you choices.

Jessie realizes she has seen this happen many times. Now she knows there is scientific grounding for why it is a good plan for her to try to help the situation by actively including someone or talking to them afterwards and drawing them in socially.

PROJECT ARISTOTLE

'Psychological safety was by far the most important of the five key dynamics we found. It's the underpinning of the other four,' says Julia Rozovsky in 'The five keys to a successful Google team'. [22] 'On Google's Re:Work website that shares their research and ideas you can see the five characteristics of enhanced teams they found.

1. Dependability – can rely on team members to get things done on time and meeting expectations

2. Structure and clarity – clear, well-defined roles

3. Meaning – personal significance to each team member

4. Impact – everyone believes their work positively impacts the greater good

5. Psychological safety – people feel safe to take risks, speak up and ask questions.

Interestingly from the perspective of the brain, there is some overlap between some of these five concepts.

Not being able to depend on colleagues can create a lower trust environment. Predictably, can also activate the threat response. You'll recall this means:

- Our brain uses up more oxygen and glucose.

- Our working memory capacity decreases.

- This impairs our creative insight, analytic thinking and problem solving ability.

- Also there can be increased cortisol release.
- This leads to decreased immunity, impairing learning and memory effects.
- Overall we experience decreased efficiency, effectiveness and productivity.

The same could be true for a lack of structure. When there is ambiguity, a lack of clarity then this can trigger the same response in the brain.

Having meaning and understanding impact, something we summarize as 'Connected to your Contribution' will likely activate the reward network in the brain.

You may remember that when people are in a reward state they:

- have better access to their cognitive resources
- are more likely to have insights which enables them to solve complex problems
- generally are more creative
- have more ideas on how to action things
- are better able to see a wider perceptual view

THE REWARD RESPONSE

Dopamine will be flowing freely. It will be released by the striatum directly into the prefrontal cortex, anterior cingulated cortex and other areas. It is important to associative learning, reinforcing behaviours, focusing attention, influencing decision-making and can elicit positive emotions.

They are going to be better at their job! And more engaged so enjoying it more. It is fantastic when science predicts something and research on the job substantiates it, as in the case with Google.

Bold Jessie

So what can Jessie do when she finds herself in situations where she feels the environment isn't ideal for her? When her psychological safety is reduced but she still wants to give her best? How can she regain control of her brain?

The first thing to recognize is that this is never in place of groups working together to create psychological safety. That is the ideal and should be on the agenda / KPI's / feedback conversations of every team in every organization until it is consistently strong. This is for those times when you work with people who are outside your culture – who perhaps aren't as self-aware or informed as you and your team.

There are several things that Jessie could do, all with the focus of shaping expectations. Practically this could include:

- framing
- gaining permission
- discussing upfront how you will work together to get the best outcome

CAN IT REALLY WORK?

Jeremy Howick, an expert on the placebo effect from the University of Oxford, set out on a mission to see what would happen to 100 people from Blackpool, England. In this part of the country one in five people have back pain. Some formed part of a control group. The rest were all told they were taking part in a study where they would either receive a placebo or a powerful new painkiller.

The pills they took were blue- and white-striped and came in carefully labelled bottles. There was a warning on the bottle to keep out of the reach of children. Their brains were scanned too.

These people had been taking a range of strong pain medications. Some of them experienced real life limiting symptoms. Half of the people taking the pills reported a medically significant improvement in their pain in three weeks. All of them were receiving a placebo.

It was found that those people who were most 'aware' and 'open to new experience' received the most benefit from the pills. There were also anatomical brain differences, including subtle differences in the amygdala.

There is mountains of evidence that placebos can have dramatic effects on people. People used to find it very strange that a physiological change could occur when people were only consuming something with no chemically active substance. While we don't fully understand how everything works yet, we do know that our expectations are very influential on our brain.

Quite simply, when we anticipate something negative we may activate brain areas involved with negative experiences and increase anxiety. When we expect something positive it is likely the brain circuits involved with the reward mechanisms will be activated.

POWER OF EXPECTATION

In 2007 D J Scott and others looked at just how powerfully expectations affect the brain. Researchers were able to look at the brain to see the activation of nucleus accumbens dopamine during the administration of a placebo. The placebo effect was linked to the anticipated effects. In other fMRI studies similar increases in nucleus accumbens activity was seen when people expected to receive money.

The people Jessie works with will have a range of expectations. The opportunity Jessie has with her language and other forms of communication is to manage and change those expectations.

ARE ALL TECH START-UPS SUPER COOL?

Synaptic Potential did some work with a wonderful tech company in central southern-east Europe. It was your classic start up story where the founders had worked out of a garage for some years before striking it big and building a wonderful team around them.

The office and people were really distinctive. They seemed to have so many things being done really well. It did seem like one big family. People were really friendly and connected. There was a bar on the roof and people enjoyed hanging out together as well as working collaboratively. We'd been brought in to do some feedback training with everyone. The neuroscience behind how it works and how to do it best.

Understanding the context that the employees were living in was important. The rate of unemployment was high in this country. Many people had endured a lot of hardship.

During the first of 10 workshops I did what I normally do, which is to aim to create an environment where everyone feels free to speak candidly. Only then can we address the deepest issues we are facing. An employee stood up and said very bluntly, 'We can't give real feedback, in case it gets someone fired'.

How brilliant that that could be voiced! A huge opportunity for the organization. Uncovering a layer of silence, for a good reason, is very valuable.

Taking the challenge that had arisen back to the founders of the company was met by deep emotion. They didn't want to hear this because they believed all their employees trusted them. The news that there was fear hurt them.

Creating a culture of psychological safety yields an organizational environment in which employees can explore and express their full brain potential.

Each person's brain doesn't work in isolation. It functions at a social level, adjusting its functioning to adapt to the context that it finds itself in. In the workplace, that context is composed of colleagues, teams, management, and the general organizational culture. And, critically, this forms the immersive environment which determines how effectively your people can express their brain potential. For example, there is no value in a good idea if the person doesn't feel empowered to speak up for fear of insult. Creating an organizational culture which scores highly on psychological safety – in other words where there is a drive to minimize and manage negative emotional and cognitive states such as fear, blame, uncertainty, insecurity, distress in favour of states which facilitate respect, tolerance, collaboration, trust, satisfaction, enthusiasm and security – provides the context in which the value of your people's brain potential can be maximized. Fostering a psychological safe environment means your people are able to express themselves, contribute, speak out and provide feedback to the benefit of their own development, and to the benefit of the organization.

MYBW Top tips for creating psychological safety as a leader of a group

- Ask people what they think.
- Tell them stories of when you've been wrong.
- Encourage them to challenge you and provide different perspectives.
- Publicly admit to not knowing everything.

MYBW Top tips for creating psychological safety within a group for yourself

- Frame what you want to say, to gain permission.
- Discuss upfront how you will work together to get the best outcome.

MYBW Top benefits for psychological safety

- People can grow and innovate more successfully.
- Individuals will speak up with concerns, ideas, praise and more.

15
Managing people, managing brains

What people's brains wished you knew

Ben had been given the opportunity to be the lead on a new project. This meant that he would have a big workload, but would also be responsible for ensuring that he got the best out of the rest of the team. He explained to Stuart that he was nervous about juggling everything but felt that he was ready. He has been working on it for two weeks now and things seem to be going well.

At one point last week he had to make some difficult decisions. The project was for a toy company and lots of the guys on the team really wanted to go and work on site. It fell to Ben to decide who got to go. After a lot of consideration Ben chose to hold a 15-minute meeting about it and laid out his thoughts and asked for feedback. He wanted all members of the team to be congruent with the choices knowing that a win for the team overall translated to individual wins too. This seemed to work really well and all the guys were on board with the decision.

There was one woman on the team who Ben hadn't worked with before, Clare, who had been transferred from another team. Ben had been told to keep an eye on her and report back on her performance. So far she didn't seem to be doing as well as Ben would have expected, but he couldn't put his finger on what the problem was.

Ben was also mindful that he has been intending to meet socially with a couple of the guys from the group but hasn't made it happen yet. He explains that he thinks it is important to connect with people away from the office building so they all know each other that bit better and are able to be more relaxed at work with each other.

This chapter helps you learn what the brains of the people who you manage wished you knew to enable you to get superior results and that in time means a lower staff turnover and more engaged teams. The bonus is a happier, easier existence at work for everyone.

Managing people – a brain perspective

Once you realize that people have complex brains managing people effectively becomes easier. In the past we have only been able to observe the result of these complex brains at work. Now we can get inside them and actually understand what is really going on, rather than just making deductions from surface level occurrences.

Managing people well involves many things. From a brain perspective there are core concepts that are important, some of which will now be familiar to you. The concepts are important in every area of business, which is why different components of them are drawn out when we think about leadership, sales, presentations, meetings or, here, management:

- trust;
- predictability;
- fairness;
- respecting reward or threat responses;
- neuromanagement by understanding the seven brain zones.

We discuss these in turn in this chapter.

Trust

The first section of Stephen M R Covey's book, *The Speed of Trust* (2006), is entitled 'The one thing that changes everything.' After outlining all the areas of our lives that are dependent on trust the author says 'trust is not some soft, illusive quality that you either have or you don't; rather, trust is a pragmatic, tangible, actionable asset that you can create' (Covey, 2006). This introduction to trust is spot on. With the advances in science, we can get into the core of what effect trust has on people and how to work with it.

Chemicals of trust

Oxytocin is a hormone that can make people feel content, calm and secure. It can increase our ability to bond with others, decrease our levels of fear and increase our ability to trust. Various studies have shown that people who have more social interactions are less physiologically stressed. Further, these studies show that oxytocin and social support during public speaking reduce people's stress levels.

Trust is vital in building strong social connections. Many managers are either unaware of the importance of trust among the people they manage, or simply aren't sure how best to create situations systematically to increase trust strategically.

TO TRUST AND TO BE TRUSTWORTHY

Paul Zak from Claremont Graduate University and a team carried out an experiment that showed what happened when people are trusted. When participants arrived for the experiment they were given US $10 for participating and grouped in pairs, although seated separately each with a computer.

One participant was asked, by a computer, to send a sum of between US $0–$10 to the other participant in their pair. Both participants are told that whatever participant one sends will be tripled in participant two's account. Participant two is told how much participant one sent, and asked to send some amount (including $0 if they want) back.

The amount participant one sends indicates that person's level of trust and the amount participant two sends back indicates that person's trustworthiness. A second part of the experiment was carried out with random drawing of amounts to be paid to participant two. This acted as a control so the experimenters could distinguish the effect of the trust signal.

The results showed that the second group of participants had nearly twice as much oxytocin in their blood when they received the trust signal (the other person in their pair sending money) compared to when they knew it was a random allocation of money. These participants returned an average of 53 per cent of the money they received, when they knew that participant one had trusted them. When the money they received was random they only received an average of 18 per cent.

An interesting point that also came from the study is that ovulating women were statistically less trustworthy. This is presumably because they release another hormone called progesterone, which inhibits the oxytocin in their own body.

Risks of trust

When trust is broken a set of things happen in the brain. The anterior cingulated cortex, which detects conflict, alerts the amygdala to a threat. The amygdala then tells the brain's reward centres and the insula that this time there won't be a reward. The motor regions of the brain are notified (via the dorsal striatum) so that future actions are affected. Threat can be detected before you can even consciously

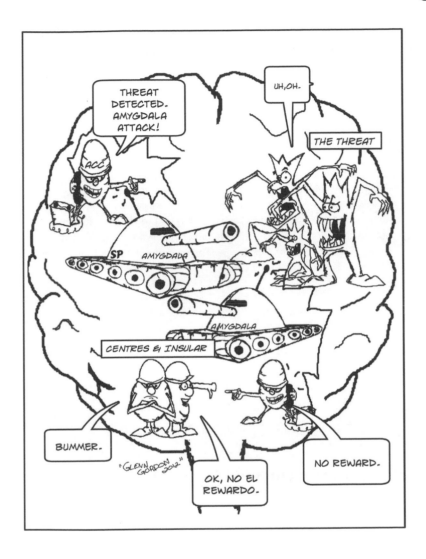

articulate the threat. While the brain is busy dealing with breaches in trust it isn't able to devote valuable resources to cognitive processes such as remembering things, solving problems and making decisions.

Working on building trust requires people to learn to read each other's intentions. As you trust people when they do certain things or meet certain conditions, your reward centres are activated. You are mobilized to act quicker, rather than being held back by suspicion or fear. Stuart broaches the possibility that Clare, the new girl in Ben's team, may be unsure yet whether she can trust the others. Unconsciously she may have picked up on Ben keeping an eye on her and be suspicious and fearful of what may happen next.

TRUST GAME

Dr Lisa DeBruine at McMaster University in Canada carried out a trust experiment. Participants were asked to play a straightforward trust game that involved them being paired up but not being able to see their partners. They had to choose between two options. The first option was them dividing a small amount of money between themselves and their partner. The second option was to trust their partner to divide a far larger amount of money between them.

Before the participants made their decision each time between which option they'd like to go with they were shown a picture of the person they were told was their partner. The pictures were actually computer morphed either to look like the participant or an unknown person. What happened next was fascinating. When people believed they were playing with a partner that looked like them they trusted the partner more. Evolutionarily this makes sense as people who looked like you were likely to be related, and subsequently trustworthy.

MYBW top tips to build trust

Making everyone you work with look like you isn't easy, although uniforms could play a part in this. Here are some alternative ways to build trust:

- Adopt team values and expectations – having a clear set of these (even if the company doesn't) and working by them consistently gives everyone opportunities to demonstrate trust and trustworthiness.

- Spend quality time together – one reason team-building days can work is because you get to spend quality time with people and see how they respond to situations; you're learning to read them.

- Be transparent – let people know what you are doing and why and try and help them be involved in the process.

MYBW top benefits of trust

- Employees are more likely to stay at a company.
- While they are there, they are more likely to enjoy greater job satisfaction and be more productive.
- Trust activates reward centres in the brain and prepares the brain to leap into action. Everything is more efficient when trust is present.

Predictability and ambiguity

In work environments people tend to work best with predictability. Knowing what is coming up is important to the brain to minimize risk and the threat response.

The situation Ben is experiencing with Clare in terms of something being not quite right may be linked to a sense of ambiguity. People are more averse to ambiguity than they are to risk. We prefer knowing what a risk is than the uncertainty of not knowing. Clare may be suffering from role ambiguity where she is uncertain what she is meant to be doing in this new team. She may be unsure why she has been moved to this team and whether she will be moved back to her old team after this project, stay here or go somewhere else. If this is the case and it continues then Clare will likely have increased stress levels and be more at risk of burning out.

If there is ambiguity in anything, for example your role, then many of the choices you make will also be ambiguous, as you don't know what you are measuring them against. This leads to evaluating the risk–benefit with the potential reward. Ambiguity is very energy intensive. The brain is constantly searching for and trying to piece together information that just isn't there. As decisions are made and actions taken, the data from them is updated to try to increase certainty. This is draining for people.

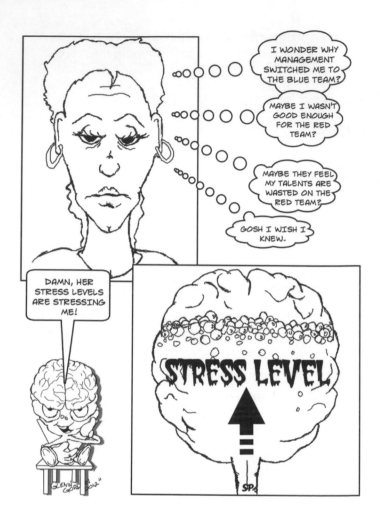

MYBW top tips to reduce ambiguity

- Elicit some clarity from whoever you can – from yourself if nothing is forthcoming from elsewhere in the company.

- Make a list of risks and rewards, get someone external to help.

- Talk in time frames and use any solid information you can.

MYBW top benefits of predictability

- Individuals' threat response is down so they are more able to think clearly.
- They are also more able to act effectively and be productive.
- Predictability helps build trust in others.

Fairness

By now Ben is familiar with the neuroscientific understanding that the brain uses the same neurological circuits for both social and physical pain and pleasure experiences. Stuart wants Ben to realize that the situation with the toy company project and who would go to work on site was handled very well in light of this.

The minefield for things to go wrong with managing people is huge. For example, if Ben had sent three out of the four people to work on site it is very possible that the remaining person would have felt left out. Social isolation can activate the same area of the brain as physical pain. This effect can be minimized if the person felt that he or she was a part of the decision-making process.

PAIN IN THE BRAIN

When you are hurt three different areas of your brain are involved:

- The somatosensory cortex – this registers where the pain is coming from.
- The insula – this tells the brain what the overall state of the body is.
- The dorsal anterior cingulated cortex (dACC) – this processes how much of a problem the pain is.

The dACC is of particular interest to us because it has been shown to be involved in social pain as well as physical. The experiment that put people into fMRI scanners so their brains could be monitored and then got them to play the cyberball game gave us great insights into a socially excluded brain. The experiment was described more fully in Chapter 10 (see page 250). The dACC in this experiment showed greater activity when the participants were excluded. When people are in pain (physical or social) it is far more difficult for them to be productive.

If any of the team members felt that the decision about who got to work at the toy factory was unfair, for example if Ben had simply chosen the two people who he had worked with the most, that is likely to have caused problems too. The social hurt of not being treated equally is very painful. The ultimatum game creates a great unfair experience for neuroscientists to study.

INSIDE UNFAIRNESS

When a fMRI study was conducted of the ultimatum game the participants played 10 rounds of the game. In keeping with the ethos of the game, each new round was with a new person.

The brains were scanned only of the responders, the people who decided whether or not to accept the offer made by the other person.

The scanner showed that participants had increased activity in the anterior insular and dACC when unfair offers were made. The anterior insular is involved in processing insult and disgust. Interestingly, the more activity the insular showed the more likely the participant would reject the offer.

A separate follow-up study highlighted another interesting facet to fairness. Participants were given drugs to create a 'low serotonin' state. In this state the participants were more likely to reject the unfair offers (the ones where their partner chose to keep the vast majority of the money, rather than splitting it more evenly). This suggests that unfairness is subjective and judgement of it can change from day to day.

Social reward

Just as people around Ben respond to his management style of minimizing their social pain, they also respond to him maximizing their social reward. Getting a group of intelligent business people together today and asking them what would reward the people they manage often turns up surprisingly similar answers: money. Money is often the first and sometimes the only answer people come up with.

INSIDE REWARD

In another ultimatum game participants' brains were studied when they were made a fair offer (one close to an equal split). It was seen that their ventral striatum (an area linked with reward) was activated when they received the fair offer, compared to when it was unfair.

A separate study looked at the comparative effect of feedback to financial reward. The participants performed two tasks, the first where they could earn money, and the second where they could receive feedback from others that varied in how positive it was. Both when they received money and when they received positive feedback their ventral striatum became active.

Receiving money is only one way of stimulating our reward centres. Ben's natural coaching style with his colleagues means that he often gives them sincere words of encouragement which really spur them on and make it easy for them to feel valued. This is excellent and something Ben should continue to do to get the most out of his team.

Neuromanagement

As you know, our brains are complex. Realizing that and understanding a little about their complexities stands you in the best position to work with your own and others' complex brains. The challenge many people responsible for managing other people face is that they are busy and so resort to presuming that other people work in the same way that they do. In any given moment an individual could benefit from your management style, but this could also not be the case.

The Synaptic Potential Neuromanagement Programme looks at the various major parts of the brain and how to best work with each of them. Stuart feels that although Ben has a lot of natural management capability, and has been on several courses, it would benefit him to understand some of these components.

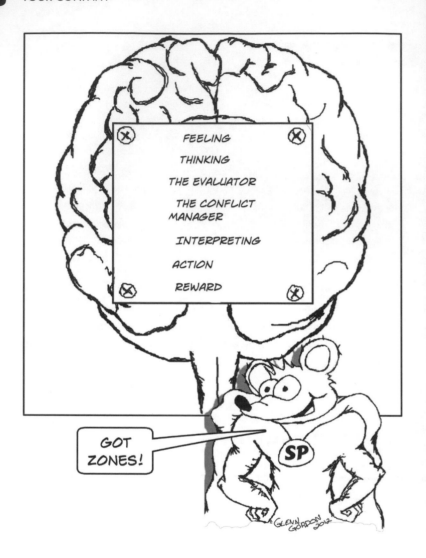

We can think of the brain as having seven major zones:

- feeling;
- thinking;
- the evaluator;
- the conflict manage;
- interpreting;

- action;
- reward.

Feeling

The parts of our brain responsible for feeling are some of the better recognized. Many managers and companies now wouldn't expect you to be able to dissociate your feelings and emotions from your daily role. A few are a step further ahead and they understand that emotions can be hugely useful to you. Emotions are key in decision making, motivation, engagement, and ultimately in behaviour. It isn't possible for people to switch off all the parts of their brain involved in feeling and still function at work.

There are a couple of major brain areas to be aware of. The first is called the hippocampus and is involved in long-term memory. It influences feeling because the memories may be emotional. People, teams and organizations have histories and it can be useful to connect with them when working on the present and future. Here are a few things that can be useful to bear in mind to manage a happy hippocampus:

- History snapshot – ask any members of a new team to explain a little of the background of their team and even company.

- Solutions snapshot – if there is a problem, get the background of how it would have been dealt with in the past and how people felt about that.

- Emotional snapshot – look at what is at stake emotionally for a person. Emotional memory can have a powerful influence on the present, and normally this goes unrecognized by managers. People can have emotions around anything; change is quite a common one.

The second very important area is called the amygdala. This region processes emotions in order of significance. Fear is top dog here and trumps any other emotion. Being very anxious can lead to an overactive amygdala. The amygdala can be activated by conscious or unconscious fear; a threat can be real or perceived. Here are a few things that will help you keep on top of amygdala activation:

- Have a plan – knowing you have a plan for multiple eventualities will reduce fear levels.
- Deal with any challenges – the amygdala responses of one person can ripple out to affect others so nipping it as soon as possible is beneficial.
- Find ways to be optimistic – authentic optimism can help to reduce fear and enable the amygdala to process other emotions.
- Change perspective – as a manager, empathizing with team members in pain can lead to over activating your own amygdala as you relate closely to them. Sympathizing can be more useful for you overall as you remain detached so you are still able to access strong cognitive resources.

The feeling brain is important in every area of business. We've looked at a few of the management applications and there are ways that a company uses the feeling brain to maximize the great experience its customers receive.

Thinking

The ability to think is probably the biggest asset of people in most jobs today. Your success is dependent on your ability to think effectively and efficiently. What you do comes as a result of the decisions you make, which are based on the thoughts you have. Sometimes we find that we aren't thinking as clearly or productively as we'd like to be.

The evaluator

There are four main lobes in the brain, one of which is at the front and so aptly named 'frontal lobe'. It contains a lot of dopamine-sensitive neurons, which are involved in attention, motivation, reward and planning.

Situated in the frontal lobe is an area of the brain that receives information from lots of other brain areas and processes it to provide a report to the action centre. You have met this area before; it is a part of the prefrontal cortex (the ventromedial part). It is known for processing risk, fear and being involved in decision making.

The challenges that managers need to be aware of with this area of the brain are:

- Anxious people tend to put too much weight on risks as opposed to rewards. This may lead to them erring too far on the side of caution.

- On the other hand, sometimes people don't register the risks enough and bury their head in the sand oblivious to warning signals all around them.

- Lonely people often don't register rewards very well.

Things that you can do to work with people in these situations are:

- Look at things from a different time perspective – either take a longer-term view of things and look at the big picture or focus on what is happening now.

- If risk assessment is poor then be mindful about assessing risks. Get help from an outsider who may be able to see things the individual cannot at that point.

- Ensure that people you are going to reward are engaged as part of a group first; giving an isolated person a reward is likely to be a waste.

The conflict manager

Another vital area of the brain is the anterior cingulated cortex (ACC). This acts like the brain's conflict manager. It connects with both the unconscious and conscious enabling it to often raise conscious awareness of important things. Ben has personal experience of his ACC being activated when he considers whether he should spend an evening with friends he doesn't get to see very often or spend the evening with his wife. He finds that even just having those thoughts buzzing around in the background slows his productivity right down. It is important for Ben to be aware that it is exactly the ACC activation that is slowing him down so that he can accurately solve the problem. The ACC will quieten down when he resolves the inner conflict and decides what his plan of action for later will be. Thinking that 'reflecting on it' will help, may do, but at the cost of some productivity.

Whenever Ben gets a feeling that someone may be experiencing some inner conflict he is wise to look into it further. Often this starts with a discussion with Stuart because this is a safe environment where he can explore what he may be noticing. Regularly, the next step is talking to the person in question, who may, or more likely may not, be aware of what the problem is. Ben then has to put on his coaching hat himself and explore things with the individual. Left unexplored things can build up and cause major problems in productivity and team dynamics.

An important caveat here is that those coaching meetings are not allowed to turn into problem-focused talks. Exploring can involve healthy framing. An overactive ACC may calm down simply by refocusing on positive things. This can be looking at an action plan of things that get done in a step-by-step way. It can be anything that reassures and builds trust because this will calm the anxious amygdala down.

Interpreting

The areas of the brain involved in interpreting are slightly less well known on the whole. Some managers would say that 'gut instinct' is airy-fairy and subsequently dismiss it. Unfortunately that would limit a person to only working with what they are consciously aware of. This is only part of the story. Unconsciously you pick up on many more things and these need processing and interpreting.

The insular is responsible for translating those internal sensations into something we can consciously make sense of. Ignoring these sensations before giving yourself a chance to interpret them means opportunities are lost. Ben had a feeling that something wasn't quite right with Clare. If he hadn't explored it in his coaching session then it could just have resulted in Clare leaving. This is what often happens in organizations. People are fearful of discussing things they don't have a complete conscious grip of and so avoid it.

Instead, it is possible to facilitate the interpreting process so that you increase your chances of acting on what you are picking up. Here are a few things you can do to help:

- When you think you have a gut feeling about something, label it as a possibility.
- Consider (list if possible) the rational reasons why this possibility may be true.
- Give each reason a probability of it occurring or being true.
- Gather more data on the possibility or test it.
- Reconsider using your additional data, conclude and take appropriate action.
- By making the process conscious you will feel confident that you can act on facts rather than just hunches.
- There is some evidence that doing yoga and Pilates can help increase your personal awareness which may help you pick up on more things that you can then evaluate.

Action

Action is quite an obvious important component of any strategy. Unfortunately, it is a part that is often neglected. When people are busy and have competing tasks vying for their attention some tasks invariably move down the priority list. Many people cite procrastination as a major challenge that holds them back from achieving.

Ben realizes there are things he has been meaning to do that are important to him that he just hasn't made happen yet. Stuart explores who Ben hasn't yet invited out for coffee and it transpires that they are the people Ben is slightly apprehensive about spending time one on one with. It is often the case that emotional barriers, uneasiness, fear or uncertainty prevent action. Waiting until he feels positive about the situation is unlikely to lead to his desired outcome anytime soon.

The best way to move things forward is often to take action. In Ben's situation he can ask himself if there is an activity he could invite these individuals to that may be more appropriate than coffee. For example, would they relax more with a beer or by helping him with a task, such as advising him on which new camera to buy? If so, he should do one of those instead, but definitely take some action.

Some great ways to get action happening are to:

- Give things a trial – act on your decided goals and test the rest. You often don't know how things will work and taking action is an opportunity to gather more data. You can then decide whether it is something you wish to continue doing. For it to work properly you have to completely commit to the trial though.

- Break the goal down into small steps – small steps decrease resistance, are less overwhelming, and each time you do accomplish a small step your commitment increases.

- Check for any bigger conflicts – there may be a deeper reason that needs paying attention to as to why action is being avoided.

Reward

The reward experience is common to everyone. You most likely know it in the very overt format of feeling excited or exhilarated when something good has happened. The ventral striatum and nucleus accumbens (NA) are actually involved in more subtle ways too. The NA is responsible for registering pleasure and reward. This means that it can be useful to activate it when we want to be motivated or to learn from positive feedback.

Here are some things to be aware of:

- Under certain circumstances, such as when people are anxious or lonely, they may not register rewards.
- Low motivation can be a symptom of a lack of stimulation in the reward brain.
- If anyone appears not to be registering rewards then the situation needs addressing quickly.

The classic ways managers' work with the reward brain is to offer short-term incentives such as money. This can be very difficult to get right on a long-term basis, although will deliver a short-term reward 'hit' to people. Another method many companies advocate is the use of social reward. Having an 'employee of the month' or peer nomination scheme will also deliver a reward 'hit' to people. The challenge with these schemes is often keeping them fresh and people engaged in them. The other challenge is there is often one, or only a few, 'winners'.

An approach that is most holistic and has the most widespread benefits is the tailored approach. Creating a company where all individuals are connected to why they are doing what they are doing is really powerful. Each time they do that thing it is possible for them to activate their reward brain. This will increase their efficiency and productivity.

Ben knows that one of his team, Natalie, is really passionate about families and being involved in local communities. When they first started working together he went to lunch with her one day with the intention of finding out more about her passion and involvement with things. That lunch was hugely valuable to him as her manager because he got to really see what lit her up. At a later team meeting he floated the idea to the team of them, as a group, supporting a local family charity and asked Natalie to suggest one based on her expertise in the area. She lit up then, and has since taken full ownership of that side of the team's existence. She updates team members each month with how much money they've raised and each quarter organizes something for them to volunteer to do outside work.

Natalie's reward circuits get fired off each time they complete a project because she knows that another small amount of money will be going to a cause she believes in and that means a huge amount to her. When she is having a bad day, or week, she reminds herself of the families the team has helped and she feels uplifted.

This strategy wouldn't by any means work for everyone in Ben's team. The key is to finding what works for each individual. That initial groundwork pays off many times over. The connection everyone has to his or her work and company is far greater than it could ever be with a few bonuses.

MYBW top tips for neuromanagement

- Embed a culture of trust proactively.
- Create some sense of predictability in people's experience of you and the organization.
- Actively consider the seven zones of the brain: feeling, thinking, evaluating, managing conflict, interpreting, taking action and experiencing reward.
- Embed these systematically into the way your company and team work.

MYBW top benefits for mastering neuromanagement

- Teams are more engaged.
- Teams are more able to produce results.
- Staff turnover rates will be lower once you have 'the right people on the bus'.

Notes

1 https://www.cbsnews.com/news/inside-google-workplaces-from-perks-to-nap-pods (archived at https://perma.cc/5LFL-J999)

2 Erin M Shackell and Lionel G Standing (2007) Mind Over Matter: Mental Training Increases Physical Strength, Bishop's University https://www.researchgate.net/publication/241603526_Mind_Over_Matter_Mental_Training_Increases_Physical_Strength (archived at https://perma.cc/2NAP-QCH9)

3 https://www.ncbi.nlm.nih.gov/pmc/articles/PMC3699874 (archived at https://perma.cc/P5ZL-Z2WT)

4 http://ei.yale.edu/wp-content/uploads/2013/09/pub184_Brackett_Rivers_Salovey_2011_Compass-1.pdf (archived at https://perma.cc/DAM9-2JJL)

5 https://edition.cnn.com/2018/04/29/health/brain-on-jazz-improvisation-improv/index.html (archived at https://perma.cc/CX2Q-S752)

6 https://thinkjarcollective.com/articles/put-your-creative-brain-to-work (archived at https://perma.cc/5CJF-S2AA)

7 Tony McCaffrey (2012)` University of Massachusetts Amherst

8 http://citeseerx.ist.psu.edu/viewdoc/download?doi=10.1.1.698.9145&rep=rep1&type=pdf (archived at https://perma.cc/R593-73DQ)

9 https://www.bbc.co.uk/science/0/21660191 (archived at https://perma.cc/LP7T-FE59)

10 https://www.sciencedirect.com/science/article/pii/S0022103109001267 (archived at https://perma.cc/JK55-82M5)

11 https://www.ncbi.nlm.nih.gov/pmc/articles/PMC4472024 (archived at https://perma.cc/7T2J-BT46)

12 S H Carson, J B Peterson and D M Higgins (2003). Decreased Latent Inhibition Is Associated With Increased Creative Achievement in High-Functioning Individuals, *Journal of Personality and Social Psychology*, 85(3), 499–506

13 https://www.scientificamerican.com/article/unstructured-play-is-critical-to-child-development (archived at https://perma.cc/YCA5-FU2B)

14 Melinda Wenner (2009) The serious need for play, *Scientific American Mind*

15 https://www.theguardian.com/us-news/2017/jul/30/marian-diamond-neuroscientist-dead-albert-einstein-brain (archived at https://perma.cc/GGE7-K2TS)

16 https://www.lego.com/en-us/seriousplay (archived at https://perma.cc/5DB9-AHBA)

17 https://phys.org/news/2018-01-lego-key-productive-business.html (archived at https://perma.cc/YE6J-4F34)

18 Amy Edmondson used these seven survey items in her dissertation

19 Amy Edmondson (2018) *The Fearless Organization: Creating psychological safety in the workplace for learning, innovation, and growth*, Wiley

20 http://fortune.com/inside-volkswagen-emissions-scandal (archived at https://perma.cc/5J2V-6SBC)

21 https://www.ncbi.nlm.nih.gov/pubmed/14551436 (archived at https://perma.cc/5X2L-YZVN)

22 https://rework.withgoogle.com/blog/five-keys-to-a-successful-google-team (archived at https://perma.cc/EHQ8-4QQZ)

Further Reading

Brodie, R (2009) *The Virus of the Mind: The revolutionary new science of the meme and how it can help you*, Hay House, London

Collins, J and Porras, J (1994) *Built to Last: Successful habits of visionary companies*, HarperCollins, New York

Covey, S M R (2006) *The Speed of Trust: The one thing that changes everything*, Simon & Schuster, New York

Covey, S R (1989) *The Seven Habits of Highly Effective People*, Simon & Schuster, New York

Cuddy, A, Fiske, S and Glick, P (2004) When professionals become mothers, warmth doesn't cut the ice, *Journal of Social Issues*, 60 (4), pp 701–718

Dawkins, R (1976) *The Selfish Gene*, Oxford University Press, Oxford

Deporter, B, Reardon, M and Singer-Nourie, S (1999) *Quantum Teaching: Orchestrating student success*, Allyn & Bacon, Boston MA

Dilts, R (2003) *From Coach to Awakener*, Meta Publications, Capitola CA

Dooley, R (2012) *Brainfluence: 100 ways to persuade and convince consumers with neuromarketing*, John Wiley & Son, Hoboken NJ

Frankel, V (1985 edition; originally published 1946) *Man's Search for Meaning*, Simon & Schuster, New York

Hsieh, T (2012) *Delivering Happiness: A path to profits, passion and purpose*, Round Table Companies, Mundelein ILL

Lehrer, J (2009) *The Decisive Moment: How the brain makes up its mind*, Canongate Books, Edinburgh

O'Keefe, J and Nadel, L (1978) *The Hippocampus as a Cognitive Map*, Oxford University Press, Oxford

Renvoisé, P and Morin, C (2007) *Neuromarketing: Understanding the buy buttons in your customer's brain*, Thomas Nelson, Nashville TN

Ries, A and Trout, J (2009) *Positioning*, McGraw-Hill, New York